HOW TO
ASTRONAUT

AN INSIDER'S GUIDE
TO LEAVING PLANET EARTH

TERRY VIRTS

WORKMAN PUBLISHING | NEW YORK

Copyright © 2020 by Terry Virts

Library of Congress Cataloging-in-Publication Data is available.
ISBN 978-1-5235-0961-4

Design by Janet Vicario

Workman books are available at special discounts when purchased in bulk for premiums and sales promotions as well as for fund-raising or educational use. Special editions or book excerpts can also be created to specification. For details, contact the Special Sales Director at the address below or send an email to specialmarkets@workman.com.

Workman Publishing Co., Inc.
225 Varick Street
New York, NY 10014-4381
workman.com

WORKMAN is a registered trademark of Workman Publishing Co., Inc.

Printed in the United States of America
First printing August 2020

10 9 8 7 6 5 4 3 2 1

CONTENTS

ORBIT

Celebrating 100 days in space during Expedition 42.

NOT YOUR FATHER'S ASTRONAUT BOOK

But He'll Like It Too!

This book is a collection of essays, each one on a different subject related to spaceflight. Some short, some long. Some technical, some emotional. Some fact-based, and some purely speculative. Some funny, some tragic. All are written with two goals in mind: to make you laugh, and to make you say, "Wow!" Often.

Many are what you would expect in a book like this. How do you prepare to handle rocket emergencies? How (and why) do you fly jets down here on Earth? Train to be a Crew Medical Officer? Perform all manner of science experiments? Prepare to go outside on a spacewalk? Make a movie in space? Rendezvous two spaceships in orbit? Take a shower in weightlessness?

Planning to leave Earth as a space tourist in the near future? There's a chapter with all you need to know.

Some essays are hypothetical. What would you do if your rocket engine didn't light to bring you back to Earth, and you were stuck in space? Hint: You'd have the rest of your life to figure it out. What would you do with the body of a crewmate who passed away? How would you handle tragic news from Earth, or bad news from bad bosses back home? Have people ever had sex in space? I also delve into the most philosophical questions of our time, about God, aliens, time travel, and how to unpack a cargo ship.

One particular chapter in this book was not a part of the original plan. But the editing was finishing up as the COVID-19 global pandemic reached full force, so I added a chapter about surviving isolation in space. It attempts to make a humorous comparison between being stuck in space and being isolated on Earth. The virus has caused so much pain and disruption to our planet, and I hope that this lighthearted take on quarantine and isolation can bring a smile during a time of unprecedented tragedy.

Although this book's subtitle is *An Insider's Guide to Leaving Planet Earth*, there are a few questions you may still have when you're finished reading it. Which country (or company) will be first on Mars? Is there alcohol in space? How much money do astronauts make and what kind of cars do they drive? Can you play *Fortnight* in space? Are there guns in space? Will the Orioles ever win the World Series again? So many mysteries.

How to Astronaut is a book about adventure. About exploration. About the unknown. It's about the best things that make us human, and a few things that make us wish we weren't. You will come away knowing more about space travel than you knew before you picked this copy up.

I wrote from the heart, in a down-to-earth style. You do not need to be a rocket scientist to digest the concepts here; they are all written so as to require no special knowledge of anything technical, other than curiosity and the desire to learn. This is not a technical manual or book of procedures. I pride myself on not explaining the precise wording of the myriad NASA acronyms you will run across (i.e., "ARED is the NASA acronym for workout machine" is about as technical as I'll get).

It's not your typical astronaut fare. But as the saying goes, I'm not your typical astronaut.

To put this book into context, here is a brief description of my career. At the age of seventeen, I left home for the US Air Force Academy, and after graduation at age twenty-one, I began my journey as a jet pilot. I first flew F-16s as an operational pilot in the United States, Korea, and Germany and finally as a test pilot at Edwards Air Force Base in California. After being selected by NASA as a shuttle pilot, I flew on *Endeavour* in February 2010 for the International Space Station (ISS) final assembly mission, STS-130. We installed the Node 3 Tranquility module as well as the Cupola, a seven-windowed observational module. A few years later, in November 2014, I launched on a Russian Soyuz rocket out of the Baikonur Cosmodrome in Kazakhstan, from the same launch pad used by Yuri Gagarin no less. After docking with the ISS, we became part of the Expedition 42 crew. A few months later, a new Soyuz arrived, replacing half of our crew, and I became commander of Expedition 43, until I returned to Earth in June 2015, 200 days after launch.

One final note: I did not use a ghostwriter. Everything here is the work of my own hand. Of course, my publisher gave me a tremendous amount of support in helping to shape and craft this work. But at the end of the day it was written by me. Good, bad, or ugly.

Godspeed (does anyone know what that actually means?) as you jump into this book. I hope many of your questions about space travel will be answered in these pages. Hopefully, others will be stimulated. Although the chapters are laid out in the order of a space mission, from training to launch to orbit to re-entry, they don't necessarily need to be read in order, so jump around if you want. My greatest desire is that you enjoy *How to Astronaut*. There won't be a quiz at the end, so just have fun reading it!

Especially if you are doing it poolside or at the beach. At least six feet away from the nearest fellow sunbather, of course.

TRAINING

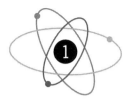

FLYING JETS

A Prelude to Flying Spaceships

Therereally is no way to completely prepare yourself for spaceflight. You can practice in simulators, study, talk to your fellow astronauts who have been there and done that. But in the end, it's impossible to prepare yourself emotionally for what is about to happen when the rocket lights and you get launched off the planet in a trail of fire, and then someone turns off the motors and therefore gravity, and you feel like you're falling (because you are).

Given this, the most important preparation I did before my first spaceflight was flying high-performance jets. Aviation is simply the closest analog we have down here on Earth to prepare astronauts for the rigors of spaceflight. It's not because of the stick-and-rudder skills of landing, or doing aerial acrobatics, or flying in formation. It's because of the mental aspects of flying—maintaining situational awareness, staying calm under pressure, making sound decisions in time-critical scenarios, staying "ahead of the jet" mentally, and anticipating several maneuvers into the future, all while zooming along at 500 mph, gas level falling by the minute, with thunderstorms bearing down on your landing airfield.

The ability to do all of these things and remain calm while your pink body—Air Force jargon for any pilot—is on the line is the most important skill any astronaut has. It's a skill that can't be taught in a simulator, where there are no real-world consequences. Fast jets are simply the best way for astronauts to hone their steely-eyed, fighter-pilot qualities.

I began my military career flying the T-37 and then the T-38 Talon, the Air Force's basic and advanced jet trainers, before going on to fly the F-16 Viper for ten years. So when I got to NASA, flying the T-38 again was like riding a bike, even though it had been a decade since my last flight in this training aircraft. But for some of my colleagues, who only had a small amount of time in light aircraft, the T-38 was a huge step up. NASA threw them to the wolves, teaching them the basics of airmanship in the supersonic T-38, a trial by fire. Thankfully, they now send newly hired nonpilot astronauts through an abbreviated military training program in the T-6, a basic turbo-prop training aircraft that is much slower than a T-38, where they learn the basics of airmanship and flying. These astronauts will never be T-38 aircraft commanders—they will always fly in the rear cockpit as supporting aircrew—but they play a crucial role, working with the pilot in the front cockpit to fly their missions successfully, and most importantly, training to get ready for spaceflight.

There are a few important skills for new crewmembers to learn. First, and paramount, is to sound cool on the radio. There really is nothing worse than someone sounding confused, or scared, or babbling on and on when they try to call the tower for permission to take off. The best advice I gave the new guys was to always sound slightly annoyed that you have to be bothered to even key the microphone to talk. You don't want to sound arrogant or like a total jerk, but you need to have an "OK, I've got things to do and let's get on with it" tone. Years later a Hollywood producer gave me the same advice when I was doing a voice-over for a video. He told me that I sounded too relaxed, and I needed to be annoyed to sound more cool. Also, new aircrew need to rehearse in their mind what they're going to say before actually talking. Clear and Concise with a touch of Annoyed is a good formula for success when talking on the radio. In the fighter community we used to joke that if we were about to crash, we would have to sound good, right up until impact. We had a reputation to uphold.

Compared to an F-16 or other, lesser fighters, the T-38 is pretty simple. As a shuttle pilot, I found the Talon to be about as complicated as a single shuttle system. For example, the shuttle's hydraulic system, or its computers, or its main engines each seemed to be roughly as challenging to master as the overall T-38. Still, I needed to study. Airplane systems. Normal procedures. Emergency procedures. Instrument flight rules and air traffic control proce-dures. Weather. Survival techniques. There's about a month of ground school that new guys go through, and annual refreshers for the old guys. Although it's a simple jet, there's still a lot to know.

Next comes the flying. You have to get used to strapping yourself to an ejection seat, putting on a helmet that is hard to breathe through, sitting in a 1960s-era cockpit that smells like a combination of jet fuel/dirty laundry/teenager's room, accelerating your body forward with afterburners, like stepping on the gas pedal, and getting smashed down into the seat when you turn the airplane, like driving fast around a corner, otherwise known as pulling g's. You have to be able to cover the inside of the canopy with a bag and fly based on instruments only, simulating bad weather. You have to get used to an incredibly high roll rate; the T-38, with its stubby wings, can actually do two complete rolls per second, though I was rarely inclined to do so.

You have to keep track of your gas. "Minimum fuel" is a term used to tell air traffic control that we were out of gas and needed to land ASAP. In the T-38 we used to say that we took off with minimum fuel. Those short wings don't hold any gas and those 1950s-era jet engines burn a lot of dinosaurs, so you're *aware* of—and low on—gas from the minute you take off. What's more, flying in Texas and the American South means flying around thunderstorms. Lots of them. I remember being taught as a student that there is no peacetime mission that requires flying through a thunderstorm, and based on some of the damage I've seen those monstrous storms do, I agree. However, in the summer they're everywhere, so a lot of your brain cells are taken up with avoiding them while getting to your destination with some gas in the tank.

NASA astronauts fly different T-38 missions, most involving basic navigation to an airport, usually 400 to 600 miles away, doing practice instrument approaches, landing, and then flying back to Ellington Field, our home base in Houston. There are several key criteria to evaluate when selecting which airport to fly to. My top priority was availability of good BBQ or other food—along with minor details such as weather, their ability to service T-38s with the air start unit required to start our jet engines, government contract fuel, a 7,000-foot or longer runway, etc. These out-and-back missions give astronauts a chance to work on soft skills such as crew coordination and situational awareness. Sometimes they go out and do acrobatics, which is somewhat useful to prepare for the sensations of weightlessness, though in truth nothing can fully prepare you for that. During the space-shuttle era we used to fly to Cape Canaveral or out to El Paso and the White Sands Missile Range to do practice shuttle approaches. The main aircraft for this was a modified Gulfstream G2, but we occasionally did these extreme approaches in the T-38, which was similar to a 20-degree dive bomb attack that I used to do in the F-16.

Sometimes, however, things didn't go as planned. And that's precisely what makes the T-38 so valuable; shuttle and station training in controlled simulator environments could not put your butt on the line to get the "pucker factor" up. One day when taking off at dawn, I hit a flock of birds, exploding the left engine. I circled back for an emergency landing on the remaining good engine. It was the shortest flight that I've ever recorded in my logbook—and probably the most heartbeats per minute of any flight I've ever had. I'm still thankful that in my backseat that day I had Ricky Arnold, one of the best mission specialists NASA ever had. One night when I was flying to Midland, Texas, the airport was hit with a rogue haboob dust storm, shutting the airfield down, and we had to fly 100 miles to the next nearest airport, sucking seat cushion through our sphincters and praying, "God, please let us make it to Lubbock." Guess who was my backseater on that day? His initials are RA.

My flying career is full of stories like these: almost running out of gas in the F-16 at Eglin AFB; finding a runway closed down upon arrival at Tallahassee and barely making it to Tyndall AFB; having my wingman lose his engine while flying over Iraq in the single-engine F-16; almost flying into a mountain on my first-ever LANTIRN (Low Altitude Navigation and Targeting Infrared at Night) flight—the computer saved my life at the last second; being disoriented and pulling my F-16 straight up at night (thankfully it wasn't straight down—I'd rather be lucky than good); I could go on and on. I've had plenty of close calls during my twenty-seven years of flying fast jets.

Although there is no direct correlation between high-performance aviation and living on the space station for six months, the possibility that some unexplained emergency could strike at any moment is what keeps you on your toes, and that is the best possible mental preparation for spaceflight.

That, and sounding cool on the radio.

SPEAKING RUSSIAN (ГОВОРИТЬ ПО РУССКИ)

Learning the Language of Your Crewmates

Most people who get selected to be an astronaut think they're pretty good at something. The former fighter pilots were the hotshots of their base, the medical doctors were the top of their field (like Goose in *ER*), the engineers could code better than any of the other engineers in their cubicle farm. But when you get selected as an astronaut you learn a cold, hard truth: Whatever you thought you were good at, there's someone better.

One skill where I thought I could hold my own was foreign languages. I wasn't that great, but I did live with a family in Finland as an exchange student in high school. And in college I had minored in French, spending a semester at the French Air Force Academy (*l'Ecole de l'Air*). Of all the fighter pilots I knew in the Air Force, I was the only one who had that kind of foreign language experience. We all tried to speak German while stationed at Spangdahlem Air Base, but it was absolutely terrible and no real German person could ever understand us; we made up our own words with a ghastly accent. Phrases like *das ist so* . . . (that is so) . . . or *du bist ein* . . . (you are a) . . . pretty much summed up the extent of our Deutsch vocabulary. It was so bad. Nonetheless, we did think we were funny and clever . . . and good-looking and humble too.

Then I got to NASA and realized something: There were other astronauts who spoke foreign languages better than I did. At the top of this list and in sole possession of first place was my eventual crewmate on my Expedition 43 space station mission, Samantha Cristoforetti. She spoke English, French, German, Russian, and of course her native Italian, all fluently and with practically no accent. She was like the Pope. After our mission the European Space

Agency sent her to China to learn how to fly their spaceship, and also learn Mandarin, only a year after our flight. When I asked her how learning the language was, she replied sheepishly, "Oh, it's OK, you know, so-so." Then I saw her give a TV interview in Mandarin on Chinese national television. Unbelievable. Samantha is probably the smartest person I know when it comes to languages. And there are astronauts like her in every field—science, flying, physical fitness, mechanical skills, you name it. Whatever you think you're good at, there's someone better.

> During our Soyuz rocket launch, Anton said, "Терри дай мне блок управления," or "Terry, give me the control panel." But блок (*block*) sounds like сок (sok, or *juice*). So I gave him a small box of juice from our food rations.

When the space shuttle program ended, we were given two options: a) learn Russian so we could fly on their spaceship, or b) find another line of work. Thinking I was pretty decent when it came to languages, I thought, "How hard can this be?" The short answer is, pretty hard.

Many people think Russian is difficult because of the Cyrillic alphabet, but honestly that really wasn't a problem. My very first day of Russian-language training was with my instructor and lifelong friend, Waclaw Mucha (pronounced *vatslov mooha*). At the end of that four-hour class, I had the letters down. It was the next fifteen years of learning nouns and verbs and adjectives and cases that would be much more challenging.

Very few words in Russian match words in English, otherwise known as cognates, though there are a few between Russian and French. For example, the word for beach is the same, *plage* in French, пляж in Russian. However, some Russian words can be ridiculously long and hard to pronounce. Hello is здравствуйте. That word literally took me five minutes to write out properly. There's no way to get around the fact that Russian's Slavic roots make it a tough language to learn because it has very few similarities to English, unlike other Romance or Anglo-Saxon languages.

The first few years studying Russian were especially painful. I had a full-time job and never studied outside of class with Waclaw. My fellow astronaut classmates and I tortured poor Waclaw, asking him to repeat the same word over and over and over in every class. I am sure that I went many months, taking a few hours of lessons each week, learning only a few new words. He had the patience of a saint to put up with my slow learning and Teflon brain. Russian words just never seemed to stick. Finally, after a few years of torture, I had reached a decent level. I learned the six cases (don't ask, just trust me, if you weren't born and raised in Russia you'll never quite get them), learned the twenty-one ways to say the word *one*, and got to a point where it was actually fun to speak Russian. Eventually, I could watch TV programs and movies—with Waclaw helping me understand each line, getting through five or ten minutes of a show during an hour-and-a-half lesson.

Getting over that hump took years, but then Russian lessons became a lot more fun for both me and poor Waclaw. It was also very important for me to be able to communicate with all of my crewmates in their native language. The Russians cosmonauts I flew with all spoke English very well, better than my Russian, but I took it as a point of pride to be able to get along in their language.

Beyond simply knowing technical language, I tried to learn cultural idioms and expressions and, more important, how to toast in Russian. This was a skill needed at the end of every major training milestone in Star City, our training base near Moscow. Being able to say a few sentences that everyone could understand went a *long* way in building our international friendship, especially during the very tense years of 2012–2015. I learned several very important skills when it came to toasting. First, when it's time to drink, don't completely empty your glass each time. This is especially important for vodka novices. Second, don't go first. Your toast will be much funnier if you are the fifth person to toast rather than the first!

There are still plenty of opportunities to mess up. During our Soyuz rocket launch, Anton said, "Терри дай мне блок упрабления," or "Terry, give me the control panel." But блок (*block*) sounds like сок (sok, or *juice*). So I gave him a small box of juice from our food rations. We almost died laughing! This story pretty much sums up my Russian-language skills. I can get along and have a conversation and make friends, but speaking for any length of time will quickly get to something I don't understand. As long as I can clarify it,

I'm fine. But sometimes when you ask for the controls of the spaceship you might get a box of juice instead.

One of my favorite things to do on the space station was to float down to the Russian segment on Friday evenings and hang out with those guys, after the work week was done. We would eat dinner, watch TV, and laugh. We also began a tradition we called "cultural program." My cosmonaut buddies would teach me expressions that I never learned in class, which provided hours of entertainment. I still remember Anton and Gennady and Misha and Sasha laughing while I learned words you don't find in textbooks (or mixed company). When we had a good satellite connection, I would use the station's limited telephone system to call Waclaw, who was usually driving home from the Johnson Space Center in Houston after work by then. "Waclaw, what does xxxxx mean?" He would always crack up, occasionally needing to pull over because he was laughing so hard at what my comrades had taught me. Those Friday evening "cultural programs" were a highlight of my time in space. One expression that my cosmonaut crewmates Anton and Gennady taught me became permanently memorialized on the bulkhead of one of the ISS modules with a Sharpie. We had a mini-ceremony as we penned this expression as motivation for future crews, laughing so hard that we would have fallen over had we not been floating. I can't repeat it in mixed company, but it basically went something like "They're hosing us, but we're getting stronger." I'll let you modify the verb.

If you are in an international environment at work or at home, making the effort to learn each other's language and culture goes a long way, and can help form a lasting bond, even when there are very strong negative external factors threatening the relationship. I'm proud to say that the crew of Expedition 43 was an example of how people from different cultures can get along, work together, and become lifelong friends, even in the face of adversity.

PAPER BAGS

Learning Not to Breathe Too Much CO_2

There are a lot of reasons to appreciate Earth. A lot. We have air to breathe and water to drink. Food in abundance. We are protected from cosmic radiation by our planet's magnetic field. There are about a million laws of nature that are perfectly tuned to make life possible here. Everything about this planet is quite amazing when you think about it.

One of those things is our atmosphere, and its cycle of O_2 (oxygen) and CO_2 (carbon dioxide). Simply put, animals use O_2 and make CO_2, and then trees and plants use that CO_2 and make O_2. What an amazing and perfectly designed life-support system. But when it comes to spaceships, meeting that need for a livable atmosphere requires a huge amount of effort by spacecraft designers, and managing CO_2 is a crucial part of that system. Because astronauts make CO_2 and there are no trees naturally on board a spaceship to remove it, a functional CO_2 removal system is mandatory to prevent the crew from dying within hours.

Because those man-made machines sometimes break down, every astronaut must learn their own personal symptoms for what CO_2 exposure feels like. If you were to breathe in too much carbon dioxide, you needed to be aware of it and act before it's too late. Because of this need for training, NASA came up with a very high-tech, precise system for each new astronaut to learn their individual symptoms.

They put bags over our faces and had us breathe in them until the CO_2 built up and we felt dizzy. Yup, I'm not kidding, that's how I learned what too much CO_2 feels like. It worked very well for us, but please don't try this at home. Under the supervision of our flight surgeon, my STS-130 crew and I all sat around a table, each breathing into a paper lunch bag, staring at each other with steely-eyed determination to outdo the guy next to us. The flight docs admonished, "Hey, this isn't a competition; just breathe until you feel your symptoms and then be done. There's no prize for going longest." Except we're all astronauts so of course it's a competition. Between pilot and mission specialist, Air Force and Marines, commander and pilot, there's always a competition whenever an outcome can be measured!

There we sat staring at each other, eyes glinting, sweat beading on our foreheads, cheeks turning purple, lips blue, eyes darting from one to the next, just exactly like when Alan Shepard and John Glenn and Gordo Cooper calmly stared each other down during their lung capacity test in *The Right Stuff*. One by one my crewmates pulled the paper bags off their faces, taking a deep gulp of beautiful oxygen. Flight surgeons pleading. Until finally there was only one of us with his face in the bag. I won't say who the last man standing was, but his initials are TV. Not that anyone was counting....

CO_2 actually became an issue for me several times in space. On my shuttle flight, we had been docked to the International Space Station (ISS) for ten days, and when it was time to leave, our whole STS-130 crew crowded back onto *Endeavour*'s cramped flight deck all at once and closed the hatch between shuttle and station. Suddenly, six people were all breathing out CO_2 in the relatively small volume of the shuttle, causing the CO_2 level to rapidly rise, and within minutes we all felt our individual symptoms—increased heart and breathing rate, flushed face, tingling lips and fingertips, stuffiness, headache. We joked that it was like being at a NASA meeting at the Johnson Space Center. Houston directed us to install an extra CO_2 scrubber (a can of lithium hydroxide that removes CO_2 from the air), and almost immediately the symptoms subsided.

During my long-duration flight, the problem of CO_2 exposure was more insidious. Our crew size varied between three and six people, depending on the Soyuz rotation schedule, and you could notice the difference when there were three versus six people on board. The CO_2 scrubbers did much better when there were fewer bodies making carbon dioxide. Those scrubbers, a combination of American and Russian machines, routinely kept CO_2 levels around 3 mm Hg of partial pressure, a level more than ten times as high as it is on Earth. On the long-duration mission it was as though we were frogs slowly boiling in a pot of water, because the CO_2 level never abruptly changed as it did on *Endeavour*—we were just continuously subjected to a slowly changing, yet very high, CO_2 soup.

It's safe to say that most crewmembers experience some kind of CO_2 symptoms while on a long-duration mission—either headaches, stuffiness, irritability, or what we affectionately call "space brain," a condition that impedes your ability to think as well as on Earth. Occasionally,

> One thing I do know: I kept my head in that bag longer than anyone else. That's my story, and I'm sticking to it.

when one of the scrubbers broke or a new crew of three astronauts showed up, increasing the CO_2 concentration, some folks would experience their CO_2 symptoms. One night, I was not able to sleep in my usual sleeping quarters because of a technical problem and had to find a place on the ISS to camp out. I picked the PMM, our storage module. After setting up my sleeping bag and closing my eyes, I began to feel my heart race and lips tingle—my CO_2 symptoms. I realized the ventilation in that module wasn't good enough, and after relocating my sleeping bag several times, I gave up and moved out into the main hallway, Node 1. It was a poignant lesson in the importance of ventilation, because in space there's no atmospheric circulation without electric fans. Without ventilation an astronaut would create a cloud of CO_2 as he breathed, and unless he moved, he would slowly die.

During my mission, we had an emergency that required the crew to use oxygen masks, and when the emergency was resolved we stored the half-empty oxygen bottles, awaiting instructions from Houston for their disposal. We were all elated when they told us to just release the oxygen into the atmosphere; the whole crew lined up to each take a hit from the O_2 bottle! You see, that oxygen was an immediate relief from our daily overdose of CO_2, and I felt a wave of relief in my brain, while also noticing my vision get a little brighter. Similarly, the spacesuit used for spacewalking has a 100 percent O_2 atmosphere. So when you go outside you get a few hours in a much cleaner atmosphere. Though there are many physical demands related to spacewalking, I really enjoyed breathing pure oxygen, if only for a few hours.

No NASA doctor really understands the long-term effects of this level of carbon dioxide exposure because people on Earth are not exposed to anything like it for the length of time that astronauts are, and there isn't a valid, comprehensive, long-term assessment of astronaut health. There are hypotheses that it could affect vision, cardiovascular health, brain function (early-onset Alzheimer's?), or who knows what else. In any case, the number of astronauts who fly is so small that most medical studies lack statistical significance.

One thing I do know: I kept my head in that bag longer than anyone else. That's my story, and I'm sticking to it.

THE VOMIT COMET

The First Taste of Weightlessness

love airplanes. And I love airplane names: Falcon, Viper, Eagle, Hornet, Mustang, Lightning, Thud, Phantom. The list goes on. As Proverbs 22:1 says, "A good name is more desirable than great riches." So, when I got to NASA and was told that I would be flying in the "Vomit Comet," I was a little skeptical.

The tradition of zero-g flights goes back to the beginning of the space program, when NASA wanted to find a way to give astronauts a taste of what weightlessness would feel like. You can of course experience weightlessness for a second if you jump off your roof, but it will be a painful landing. If you have a diving board (we used to have those before we had lawyers), you will also get a second of free fall before splashing down. If you fly in a light aircraft, the pilot can push forward on the stick and get a few seconds of zero g. But to really get a significant amount of time, you need to be going fast, and it turns out that an airliner makes a perfect way to get more than twenty seconds of floating.

The mechanics of how this works are pretty straightforward. The Vomit Comet flew out to a 100-mile-long, restricted operating area over the Gulf of Mexico with plenty of airspace in which to maneuver. It would climb up to 20,000 feet and nose over, accelerating to 350 knots. At the designated speed, the pilot would pull back on the yoke, smashing all of the passengers down on the floor at two g's, or twice their normal body weight. When the nose had pitched up to 45 degrees above the horizon, he pushed forward on the yoke until the plane was at zero g's. This is a bit like being at the top of a roller coaster as it pauses at the peak of the tracks. Except it's *way* more dramatic and lasts a lot longer. It took about twenty-five seconds for the plane to float over the top of this parabolic arc, until the nose was pushed down to about

30 degrees below the horizon. At that point, the pilot firmly pulled back on the yoke, once again smashing the poor occupants of the plane against the floor at two g's. If you were a physics major, the trajectory of the airplane described a parabola.

This porpoising motion—nose down, pull back, smash everyone against the floor, push forward, make everyone float for twenty-five seconds, repeat—continued for ten parabolas. At that point, the plane reached the end of the restricted airspace, so we had to turn around and do it all again. Most of these flights involved forty parabolas during a two-to-three-hour sortie.

When I first arrived at NASA, they were still flying the KC-135 Vomit Comet, the workhorse of American zero-g flights for many decades. That plane was awesome, with a big open cabin, padded walls, lots of room for the poor occupants to float around and practice weightlessness or operate experiments that scientists were testing out before sending them to space. A few years later, in a cost-cutting move, NASA transitioned to the C-9, a smaller, two-engine aircraft, which meant lower fuel costs compared to the four-engine KC-135. This plane was actually pretty good, because it had about 90 percent of the usable volume and was cheaper than the KC-135.

I made it for about ten parabolas before tossing my cookies. There's a funny picture of me kissing the ground, surrounded by my classmates, after that flight.

A few years later, HQ had the bright idea to privatize this program. It was contracted out to a commercial company that ended up costing more money than the in-house NASA aircraft, and our scientists got less valuable data. Shortly afterward the program was shut down entirely, and now NASA doesn't have a zero-g capability. The Russian and European space agencies still have their own versions of the Vomit Comet, but NASA doesn't. Although privatization is often a good thing, in this case it wasn't. NASA's flight operations were efficient and built upon synergies with other aspects of its aviation program, and it was a shame to lose such an important part of spaceflight training in the misguided name of privatization.

My first zero-g flight was accompanied by great anticipation. I was a fighter pilot and test pilot, having flown more than forty different types of

NASA's KC–135 "Vomit Comet," where astronauts learn the thrills of zero g (and sometimes barf).

aircraft. But those parabolas, flying up and down and up and down in the steamy Houston humidity, in the back of a musty 1960s-era airplane, with the smells of thousands of prior Vomit Comet flights hanging in the air, concerned me. My ASCAN (Astronaut Candidate, the NASA acronym for new trainee) class was going to fly together, and we all decided that we would do this first flight with no motion-sickness medication; we would all experience weightlessness for the first time naturally, in solidarity. So, against the strong recommendations of our flight surgeons and other senior astronauts, I tried my first flight *au naturel*—no medication. Ugh. Those parabolas, up and down, floating and then getting smashed to the ground and floating and smashing some more, not being able to look out the window, the smells of decades of sick students and musty jet fuel, up and down—forty times. I made it for about ten parabolas before tossing my cookies. There's a funny picture of me kissing the ground, surrounded by my classmates, after that flight. If you ever decide to take one of these zero-g flights, as either an astronaut or a paying customer, I have one recommendation for you: Take the meds. I only later found out that a lot of my friends had cheated and done exactly that. They were the smart ones, for sure.

The experience of weightlessness was spectacular, though, unlike anything I had ever felt on Earth. I felt as though I were falling, with no pressure on any part of my body, nothing to stabilize me, and the slightest push on a wall or classmate would send me shooting in the opposite direction. It was also a battle dealing with the fifteen other people flailing around inside the cabin, everyone feeling an urge to kick their legs or wave their arms as they flailed about in a state of disorientation. I needed a football helmet and mouth guard to protect myself. After about thirty minutes I gained a level of confidence, having learned the basics of keeping my body under control.

Perhaps the most important thing to know about the Vomit Comet is what to do when the weightlessness ends. At the end of the parabola, the pilot calls back to the passengers floating around that it's time to pull out, and the loadmaster yells, "Feet down, pulling up!" When you heard that, it was time to get to the floor, with your body flat against the ground, because in about two seconds the plane would abruptly pull up at two g's or more. If you were still on the roof, you would be slammed to the floor. And if you were standing on your feet or arms, you would suddenly have a whole lot of body weight pushing on them, possibly leading to an orthopedic injury. I also tried to keep my head stable and upright during the pullout. This was to help with nausea, because moving your head around during the pullout supposedly messed with the vestibular system in the inner ear. I'm not sure if this was true, but I always kept my head still and it worked for me.

Getting astronauts used to floating was an invaluable mission of the zero-g program, though having them test scientific equipment in a weightless environment before sending it to space was even more important. One of the most memorable experiments I did on the Vomit Comet was testing a new intubation device, used to give a patient an artificial airway in the event of a medical emergency in space. Getting a tube down the trachea (and not the esophagus) is a very tricky thing for well-trained medical personnel to do on Earth, much less a fighter pilot like me in space. So they developed a new shoehorn device to help the tube go down the right hole behind the tongue, and I practiced the procedure on several zero-g flights. Another memorable experiment was performing ultrasound imaging of the carotid artery in my neck to investigate the effects of weightlessness on the cardiovascular system. I was stunned to see the artery narrow while under g-force and open up dramatically as soon as we got to zero g. I was thinking that I should be passing out when my major neck blood vessels were opening and closing so dramatically, but my brain kept on working just fine!

Yet another experiment involved a series of MIT-developed satellites somewhat ironically called SPHERES because they were small cubes, not much bigger than a lunchbox. They were used for testing satellite station-keeping. When the plane went to zero g, we would release them and they would fly around in formation, communicating with one another, as their larger cousins would eventually do once they got to space. It was like being on a *Star Wars* soundstage. A decade later, it was a great pleasure to fly with SPHERES satellites actually on the space station. Seeing those small robots flying around the ISS for real was like being on a *Star Wars* spaceship, not just a soundstage. When we didn't have formal NASA-sponsored experiments to test on the Vomit Comet, there were usually college students on board, flying their own university-sponsored engineering projects.

The zero-g flight program was one of NASA's most successful and useful projects for decades, giving thousands of astronauts, scientists, and students the chance to experience weightlessness and test out their hardware. It helped me mentally prepare for spaceflight and weightlessness as much as possible. I've said it before and I'll say it again—if you do this, take the meds! Or you'll be kissing the ramp after you land . . .

SURVIVAL TRAINING
Preparation for Space Calamity

have to admit, one of my least favorite parts of being in the Air Force was going through survival training. After finishing my freshman year at the Air Force Academy at the age of eighteen, I had to do SERE (pronounced *seree*) training: Survival Evasion Resistance Escape. It was a dreaded rite of passage that had come to prominence during the Vietnam War, during which America's airmen were treated horribly, often for years, as prisoners of war in the Hanoi Hilton, the nickname for the infamous Hỏa Lò prison. The military devised the SERE program as a way to give flyers the skills they would need in future conflicts to survive, evade, and resist. In theory this made sense to me, but in practice I wasn't a big fan of starving or freezing or being tortured, so I was less than enthusiastic as I went off to SERE training in July 1986. Little did I know that survival training would be an ongoing part of my professional life for the next thirty years.

I was back in the woods with the French Army as part of my exchange to the French Air Force Academy in 1988. Before I could complete Air Force pilot training, I had to do water survival in the Florida Keys. Then it was off to the mountains of Maine with the US Navy when I was selected to be an astronaut. After all of those experiences, I figured I had learned all there was to learn: how to deal with cold and wet weather, to "conserve sweat not water" in the desert, that skin-to-skin contact was crucial for preventing hypothermia, to stay dry at all costs, that you needed clean water much more than food, etc. No need for more training, right? Wrong.

NASA partners with the NOLS (National Outdoor Leadership School) on a program designed to prepare crews for the psychological rigors of extended spaceflight. There are several types of NOLS courses: hiking, sea kayaking, sailing, winter survival, etc. About five years into my astronaut career, I was assigned to the sea kayaking course in Prince William Sound, Alaska. Then seven years later, after being assigned to my second ISS flight, I went on another NOLS trip, again to Alaska. These expeditions were chances to get to know possible future crewmates under trying conditions. Little did we know just how trying the conditions would be!

The NOLS Alaska office is near the airport in Anchorage, and after a long flight followed by a short bus ride we were soon putting together bags of gear, learning the basics of sea kayaking, and getting to know our two instructors. The key to any camping trip is to pack lightly and keep everything waterproof, and kayaking was no exception. Clothes were first vacuum sealed in Ziplocs and then stuffed into thick, waterproof duffel bags. Minimizing bulk was important—both of my trips were for more than ten days, and everything had to fit in the kayak. Clothes (one spare shirt and pants), underwear (fresh pair every few days), socks (three pairs), one fleece layer, and one good Gore-Tex jacket. Food. Tents. Cooking gear. Fuel. A journal. There wasn't a lot of space for the twelve of us to cram all of that gear into our eight kayaks.

There are many analogs between NOLS trips and actual spaceflights, and packing and keeping track of gear is the first. It's also something I've struggled with my whole life. Where are the keys? Where is my wallet? Where is the camera? Where are my shoes? Losing things is a pain on Earth, and can make a camping trip miserable, but will absolutely ruin a space mission if you have to waste time looking for your stuff. Storing and managing gear in backpacks or jackets or Ziplocs is a skill I transferred from the Alaskan wild to the weightlessness of *Endeavour* and the ISS.

Sea kayaking itself was a challenge. The job of group leader rotated from astronaut to astronaut each day. A typical day's mission entailed paddling from one island to another, across a few miles of open ocean, to a new campsite. Those open-sea crossings were high-risk, because we were miles from the nearest shore, and the unpredictable weather could quickly make conditions extremely hazardous for small kayaks. The leader of the day would assemble the group, study the weather brief, select our next campsite, plot out the route, and take a go/no-go poll of the whole group. Once everyone agreed with the plan, we ate a quick breakfast, broke down camp, loaded our gear, and put out to sea in our kayaks.

Getting the kayaks from the beach to the ocean was neither simple nor graceful. First, there were waves, which made things interesting. Second, kayaks are not exactly stable; they are easy to flip! It's like riding a telephone pole in water. We would push the kayak halfway into the water, with a bit of the tail resting on the beach to stabilize it, and then push off into the waves, quickly attaching a rubber skirt from our waist to the kayak itself, creating a waterproof seal to keep our legs and butts reasonably dry. I was very lucky to never have flipped over, but not all of my crewmates can say the same thing.

Kayaking itself was so tricky that our instructors conducted special training on our first day—egress drills. The goal was to be sure you could get out quickly in case the kayak rolled upside down. We started in the sea, at rest, bodies tightly sealed into the seat, rolled the kayak over 180 degrees so we were completely underwater, then yanked on the emergency tab to release the rubber skirt and shimmied out of the kayak. When my head finally popped out of the water, I tried to let out a scream but couldn't because I was in shock from the balmy Alaskan water. There was actually an iceberg floating next to us! After the initial shock subsided, the next task was to get back on board the kayak, because if this happened for real, in the middle of a channel, miles from the coast, there would be no swimming to shore. The biggest challenge was flipping the telephone pole (aka kayak) upright, then mounting it from the side.

> I'm sure it wasn't a big deal for the orca, but to be in the ocean, in the food chain, with this majestic, beautiful, powerful, intelligent (and hopefully not hungry) creature gliding alongside us was sublime.

After being dunked in the arctic seawater, we made our way back to shore, and it was time to dry off and warm up. For this exercise we had worn "poopy suits," unfortunately nicknamed rubber overalls designed to keep aircrew warm and dry in freezing water. Except they were ancient and therefore full of holes, so they just let the freezing water in. And kept it in. By the time we made it to shore, we were ready to dry off and get something warm to drink. Egress training was fun in a weird way, though, building camaraderie in our group and giving us some much-needed confidence that we would be able to handle those wobbly kayaks in the open Alaskan sea.

The best part of those NOLS trips was the Alaskan scenery and wildlife. The first thing that anyone venturing into the Alaskan wild needs to know is the bear situation. Is it black bear, brown bear, or white bear country? Thankfully, we were in the south, which is black bear territory. Brown bears, or grizzly bears, are an entirely different ball game, with a different set of rules and a hugely elevated risk. Simply put, they eat people. Frequently. I was told that in the event of a black bear encounter I should try to scare it away, or if that didn't work to play dead and maybe it would leave me alone. In the event of a grizzly bear attack I'd have to fight back, 'cause it wasn't going to just leave. Of course, fighting back against a giant creature with those claws and teeth . . . well, the good news is I wouldn't suffer for long.

Polar bears are even worse. They are so large and powerful that the danger was absolute if you encountered one in the wild, unprepared. In fact, a Norwegian friend of mine told me that it was illegal to be in polar bear country in Norway unarmed. They require you to carry a gun. Ironically, they also make it illegal to shoot a polar bear. I guess it's either die or go to jail. I assume Norwegian jails aren't that bad, compared to being eaten alive, but I think the best course of action is to avoid getting in that situation to begin with!

That first week in Prince William Sound we saw constant evidence of bears—scat, scratch marks on trees, and lots of half-eaten salmon, which were in peak spawning season at the end of August. One day we saw a black bear lumbering down a trail a few hundred yards away as we paddled along. That was such a powerful moment for me, seeing a bear in the wild, doing its thing, no humans anywhere nearby. About a day later we landed at a new campsite. We had been seeing small schools of salmon flopping around in the shallows for days, but at this beach there were thousands, maybe millions of those fish that had reached their final destination. I'd never seen anything like it. They'd found the mouth of a river, their birthplace, and were thrashing and flailing with every last ounce of strength left in their bodies, pushing upstream, where they would lay their eggs and die. Those were some nasty fish; their skin was shedding and they looked awful and smelled worse, using their last drop of life to make it to their own burial ground, sowing the seeds of their next generation. This was an ancient rite of passage, repeated endlessly throughout the millennia. It was an otherworldly honor to see nature in its rawest form. Life, death, rebirth, all in one location.

With all of that sushi lying helpless in a few inches of water, there were—you guessed it—bears. Lots of them. As evidenced by scratched trees and scat. We were on edge at that campsite, to say the least. When walking around alone I yelled, "Hey, bear bear bear!" to scare them away. At night, when it was time to get out of the tent at 0100 for a bathroom break, I turned several flashlights on and made a ton of noise. Our stress level lowered when we left that sacred salmon spawning ground the following day!

Living by the leave-no-trace mantra was a big part of the NOLS experience. We brought most of our own food and brought all of our trash out of the woods with us. It amazed me to see how it was possible to live with so much less stuff than we do in modern America. We even went without toilet paper. Yup. It's possible, actually even comfortable, to go (so to speak) without it. The good news is that in Alaska there are lots of smooth stones along the seashore. And thick, soft moss. So, no TP required. We packed basics—flour, eggs, sugar, butter, bacon, etc.—and we also got food from the wild. The blueberries were incredible, and the fish were amazing. One day my friend Leland Melvin and I were fishing while a school of salmon floated by, right next to the shore, fins sticking up like mini-sharks. Throwing the fishing line into that school wasn't fair, so I snagged the fish by their fins—they didn't even have to swallow the hook. To this day we still laugh about that fishing trip. Needless to say, we ate well that night, having freshly caught salmon and wild blueberries, supplemented with butter and bacon. It felt like that's what I was wired to do, living in the wild and catching my own food. Going to the neighborhood grocery store or chain restaurant to hunt and gather just doesn't satisfy those deep-seated instincts.

Cooking was a skill that I didn't possess before NOLS—I never cooked at home, so going into the woods with a camp stove and bags of flour and sugar and butter was a new experience. Luckily, we were with an Italian astronaut,

Paolo Nespoli, who showed us all how to make pizza one night. That was one of the best meals I've ever eaten! Food was a highlight on both of my NOLS trips, and I learned several important lessons. First and most important, more butter! Whatever you are cooking, it will taste better with more butter. Second, more bacon! Everything goes better with bacon, especially if you are freezing and wet in the Alaskan wilderness. Finally, finding food in nature is hard; it doesn't just appear magically. It's hard work hunting and gathering food, and survival isn't guaranteed. In most of the world, we live such incredibly sheltered lives; it's worth taking some time to live in the woods to appreciate the relative ease of our middle-class lifestyle, with grocery stores full of everything we could possibly want within minutes of our homes. Had we not brought those bags of food with us, we all would have lost twenty pounds during those short trips into Prince William Sound, even with the ample supply of blueberries and salmon. Nature can be brutal.

> The goal of those NOLS trips is to make us miserable, so that when we face difficult situations in space, we have the experience of surmounting similar situations and working with crewmates when everyone is crabby.

One of the most spectacular moments of my life happened unexpectedly. We were paddling in our group of eight kayaks, enjoying a mild, cloudy, rain-free day, surrounded by gorgeous mountains, when there was a disturbance, behind me and to the right. Out of the water popped the tremendous dorsal fin of an orca, curved over, smoothly gliding through the water parallel to our kayaks, maybe 10 meters away. I'm sure it wasn't a big deal for the orca, but to be in the ocean, in the food chain, with this majestic, beautiful, powerful, intelligent (and hopefully not hungry) creature gliding alongside us was sublime. It was a moment that helped me understand that we are a part of nature and there are incredible creatures out there, too, sharing the same land and air and water with us. Beyond those lofty thoughts, I was glad that the killer whale didn't eat me, because he certainly could have.

I also saw other wild creatures, like sea lions, otters, and seals, on those trips. There was one particular island on the south side of Prince William Sound that had hundreds of those loud, squawking pinnipeds. They were lying there on the rocks, yelling and screaming for us to go away, or maybe warning each other of impending human kayaker danger, or trying to entice a mate to join them for some afternoon action, or just trash-talking the seal tribe on the next rock over. Whatever the reason, those beasts definitely use their outdoor voice. I'd hate to have them as next-door neighbors. But they sure were cool to see in the open ocean.

Our wildlife experience wasn't limited to sea and land. There were also birds, including interesting puffins that looked like flying penguins, garden-variety falcons and hawks, and most of all bald eagles—lots of them, with their majestic and powerful black bodies and distinctive white heads. We joked that they were as numerous as the mosquitoes. I remember as a kid that seeing a bald eagle was a really big deal; they were rare, and usually seen only in zoos. But in Alaska they were everywhere. They have a very distinctive, high-pitched, shrill hunting call, which doesn't match their impressive appearance at all. The first time I saw one I was so excited, but after a few days the sightings became commonplace. I never took it for granted, though, and like a beautiful full Moon, or a sky so clear you can see the Milky Way, or a fiery red sunset, I tried to enjoy and appreciate each one.

Weather is an issue when you go sea kayaking in Prince William Sound. A big issue. My first trip there had very nice weather, which the instructors hate. The goal of those NOLS trips is to make us miserable, so that when we face difficult situations in space, we have the experience of surmounting similar situations and working with crewmates when everyone is crabby. But I couldn't complain about this trip; we had a few days of perfect, sunny weather. Then a few days of cool and cloudy. Then drizzle on the last few days, with only one nonstop-rain day. I counted myself very lucky, because the group that went after us that year had a miserable time, though I'm sure their instructors were happy.

However, when I went back a few years later with my Expedition 42/43 crew, the weather was much different. It rained every minute of our ten-day expedition, with the exception of one afternoon. The weather at our first camp was severe, with strong wind, rain, and rough seas, making kayaking out of the question. After we sat around and twiddled our thumbs for a few days, they sent the big boat back out to pick us up and move us up north to a more sheltered fjord. The sun came out for a few hours during that boat ride! But when we arrived at our new camp, back came the rain. We were able to do some paddling at the new location, in constant rain, until a few days later when a serious storm showed up. It was near the end of our planned trip and we had to decide—ride out the storm in place, leave for a new campground, or declare victory and end our expedition early, before the planned ending date.

My vote was to pack up and leave. The storm was predicted to be hurricane strength and we would essentially be hiding in our tents, a few hundred feet from the shore, hoping that a tree didn't fall on us. Most of the group wanted to stick it out, proving that we could handle adversity, so that's what we did. In the end it worked out, nobody was hurt, and we paddled back home after a few days. But this incident reinforced an important lesson I learned during my time at NASA about risk management: If you make a decision and everything works out, it doesn't mean you made the right decision. To emphasize the point, the space shuttle program had lived with foam falling off our fuel tank during launch for years and it had always worked out, until the final flight of *Columbia* when it killed the crew. And just because we rode out a hurricane in the Alaskan wilderness and survived doesn't mean it was the right call; it might just mean that we were lucky. Of course I'd rather be lucky than good, but there's an important lesson here: Debrief and analyze your decision-making not based on the outcome, but to understand if the right decision was made, given all the facts known at the time.

Sea kayaking was much more physically demanding than I had imagined. Paddling those kayaks miles across the open sea, from one island to another, was hard work. There were a few fjords that acted like wind tunnels, and one day was particularly brutal, paddling against the wind and waves. There was a mix of single- and dual-place kayaks, and I was paired with Leland Melvin for several days in a dual. It didn't take long for me to realize my good fortune. He had been an NFL wide receiver before becoming an engineer and then NASA astronaut, and he was an awesome partner for the dual kayak! I have a great picture of me sitting back and relaxing while Leland paddles away, almost kicking up a rooster tail, leaving the other experienced (i.e., old) astronauts in the dust. As they say, "Work smarter, not harder," and getting paired with Leland was the luckiest day of my brief kayaking career.

One beautiful day we were paddling down a fjord on the southwest area of our course. We noticed something big on a beach about a half mile away. After a quick poll, the group decided to go check it out. The closer we got, the more bizarre it appeared, looking like a cargo trailer from an 18-wheeler. When we got to the shore it became apparent what it was—a beached whale. That had been beached for a *long* time. As we climbed out of our kayaks and approached the beast, two things stood out. First, the smell. It was the worst smell I'd ever experienced in my life, just absolutely awful, a combination of death and rot and spoiled food and nasty, and you could smell it 100 yards away. Second, it was moving, kind of like a plate of jiggling Jell-O. As we got closer it became apparent why. Maggots. Millions of them, maybe billions. White blankets of maggots feeding on this immense carcass. The smell and overall disgusting-ness of the situation was overpowering, and we spent only a few minutes there, paying a brief tribute to this giant, formerly magnificent sea creature. I don't know what species of whale it was—definitely not an orca, because it was too huge. And we didn't know how long it had been there, but my guess was a year, because it was more than half decomposed, even though there were still many tons of biomass and bones there. The cycle of life in the wild—death, scavengers giving back to the food chain, and life—all in the overpowering beauty of Alaska. Yet another stark reminder of the harsh reality of survival.

Though we were sea kayaking, there was ice everywhere. On several occasions we camped next to glaciers that were emptying into the sea. The experience of being next to one of those glaciers is a bit intense: They are massive, and cold. Air is chilled by flowing across the ice, and being near a glacier often means that the air is 10 degrees cooler than it is just a few miles away. Most impressive is the noise of the glacier calving, dropping massive icebergs into the sea. That sound is haunting, a tortured ripping of tons of ice from the mother glacier, followed by a tremendous splashdown in the water and a prolonged, roaring thunder. This happened every few minutes. All day and night long. It was as if the glacier were alive and readying an army for battle. Of course, all of this ice falling into the ocean meant that there were icebergs everywhere. It was cool to paddle alongside a massive iceberg and tap it with my paddle; they felt as sturdy as a mountain. The ice was white on the outside but often had an intense blue on the inside. The edges of those icebergs were very sharp, and if I were a boat captain I would stay far away. Besides the noise and icebergs, there were waves. The closer we got to a glacier, the more unstable the water got, especially for small kayaks. Like bears or rocket launches, they were nice to view from a distance, but I had no desire to paddle too close to those mountains of ice crashing into the sea.

Glaciers contain millions of tons of frozen water and are continuously flowing across the land in a never-ending march to the sea, creating a constantly changing mosaic of crevasses and canyons as the ice cracks and shifts. Because of climate change, the size and quantity of ice in the Arctic has undergone dramatic reduction, and I saw this firsthand on my NOLS trips. Our instructors had photos of where the glaciers had been ten years prior, or in the '80s, or '40s, or even a hundred years ago. We would compare the old photos to a modern-day glacier, and there was no doubt that the ice had significantly retreated. When land ice melts, that water flows down into the ocean, which is a primary cause of sea-level rise. For me, it was one thing to hear about climate change on the evening news, but it was another to see it with my own eyes.

After all of those survival experiences in the military, NASA, and NOLS, I hoped I was done with formal survival training. But as they say on TV, "Wait, there's more!" My second spaceflight was on a Russian Soyuz, and part of that training was, you guessed it, survival training. Both winter and sea survival. The first one on my schedule was winter survival, which is a rite of passage for cosmonauts, but also something we knew might be needed. In 1965, the first man to do a spacewalk, Alexei Leonov, along with his commander, Pavel Belyayev, landed off course in the Ural Mountains and spent several nights in the freezing woods there. Then in 1975, a Soyuz crew aborted about five minutes into their flight and ended up landing in freezing Siberia, in the Altai Mountains. Legend has it that their capsule almost slid off a cliff until the parachute snagged on some trees, saving their lives. Then in 2018, another Soyuz rocket failed and the crew made an emergency landing, though this time they landed much closer to the launch pad—and farther from the grizzly bears. In addition to the possibility of landing in freezing Siberia, there was the possibility of unexpectedly ending up in the ocean, especially given the fact that most of Earth is covered by water. We all knew that an unplanned landing in a dangerous location was a possibility, so off we went for more survival training.

When I think back on these experiences, from my first SERE course at age eighteen to my NOLS trips and Soyuz winter and water survival training in my forties, what stands out most is being in nature. The smells of the Colorado forest. Seeing the Milky Way through the trees in Maine. The overpowering beauty of Alaska—the ocean, mountains, glaciers, orcas, eagles, salmon, bears. Wild and rugged terrain at each beach campsite while kayaking; fine sand, smooth stones, some with annoying little sand fleas, others surrounded by bald eagles. Tasting wild berries, cooking fresh salmon, setting up and breaking down campsites. Marching for hours and hours with crazy French Army sergeants yelling. Freezing in a homemade *shalash*, or shelter, in the Russian winter. Parachuting into the warm Caribbean by the Florida Keys, or floating in the canal where cooling water from the Turkey Point nuclear power plant empties into Biscayne Bay, and seeing bizarre two-headed fish and broccoli-like creatures floating along with me, thinking of Homer Simpson running the safety program for the water I was stuck in.

Spending intense time with my comrades, being pushed to our physical and mental limits, did more for my self-confidence than anything I've experienced. When I was eighteen years old, I was so immature and afraid of situations like these. I now realize how much of an impact these survival courses have had on my life.

Along the way, I learned some important life lessons. Smooth stones can be useful. Everything tastes better with bacon and butter. Be sure to ask Leland to join you if you're going on a kayaking trip. If you've never experienced nature viscerally, do it. Take some time, get a tent and a sleeping bag and a suitable companion with whom to share body heat, and go see the stars or bears (black, and only from a distance) or orcas in person. iPhone off. Trust me, it will be worth it.

6

SPACE SHUTTLE EMERGENCIES

The Special Hell Created by Simulation Supervisors

y friend and legendary space shuttle program manager and flight director, Wayne Hale, wrote a great article describing NASA's simulator supervisors. Sim Sups (pronounced soups) are the people whose job it is to make astronauts' and mission controllers' lives miserable. He used words like "nexus of evil," "diabolical," and "insidious," among other choice terms. And he was right on. During my decade as a shuttle pilot, I spent countless hours studying, training for, and debriefing on an endless litany of malfunctions that these people threw at me. Some realistic, many absurd, all intended to best prepare the crew and flight control team for whatever awaited them in space.

The process of becoming a shuttle pilot took years. When I first showed up at NASA, I had a somewhat smug attitude, thinking that the shuttle would be an easy vehicle to master. After all, I was a test pilot, and it couldn't be *that* complicated. It didn't go to the Moon like Apollo did, so how hard could it be? Boy, was I wrong. I've heard many old-timers, who worked both Apollo and shuttle, say unequivocally that the shuttle was the most complex flying machine ever invented. Having flown her, I must say that I agree.

In order to learn such a complicated beast, our trainers broke the shuttle down into its various components. There was also an entire language of acronyms that we had to learn, along with vocabulary and grammar and idioms, just like learning Russian. There was the DPS (computer system), ECLSS (life support), APU (hydraulics), EPS (electrics), RCS and OMS (thrusters), MPS (rocket engines), PDRS (robotics), etc. Each of those subsystems was about as complicated as our whole T-38 jet trainer. Worse just than having to learn each one individually, we also had to learn how they all interacted with each other, and it was in those interactions that our Sim Sups could really get devious.

For example, if an electrical component failed, it would have implications for the shuttle's main engines. As the pilot (equivalent to a copilot on an airliner; the commander was the equivalent to captain, or boss), my main job during launch was to make sure the three main engines were running, so electrical failures would definitely grab my attention. Except, after giving us an electrical malfunction, the Sim Sup would pile on ten additional malfunctions, each one having a unique interaction with another, gradually building up a doozy of a worst-case scenario. Then would come the *coup de grâce*—a second electrical malfunction that would require me to shut down a main engine manually, or else we would blow up. Of course, that shutdown would have to happen in the next thirty seconds, so the crew and mission control would have to communicate this need *very* quickly and succinctly. And oh, by the way, as soon as the main engines shut down and we were in simulated space, there would be another rocket burn required to stabilize our orbit, because those malfunctions had led to an underspeed. While we were busy getting to a safe orbit, I would also have to shut down the hydraulic pumps (did I mention that they also blow up if not shut down properly?), all while fixing the cooling system. Oh yeah—if the orbiter overheats you die, so you either get that fixed or do an emergency abort, returning to Earth before you even complete one orbit.

Although we all knew that such a combination of ridiculously well-planned malfunctions would never occur, we also knew that it was impossible to simulate the stress and "pucker factor" of an actual launch.

The shuttle was a complicated vehicle, to say the least. When these emergency simulations went well, it was truly a thing of beauty to see the intricate ballet of rocket science and *The Right Stuff* coming together to handle some of the incredibly difficult scenarios that those diabolical Sim Sups had devised. Although we all knew that such a combination of ridiculously well-planned malfunctions would never occur, we also knew that it was impossible to simulate the stress and "pucker factor" of an actual launch. Having launched on a rocket, twice, I can verify this. There is a stress that comes from riding on millions of pounds of high explosives that can't be simulated. So

NASA decided to max out our brains with these uber-challenging scenarios as a way to prepare us for the tremendous stress of launch day. I think that was a good plan.

When I was an F-16 pilot, we had a system called LANTIRN (Low Altitude Navigation and Targeting Infrared at Night). It allowed us to fly close to the ground at night, avoiding enemy surface-to-air missiles and fighter jets, while dropping laser-guided bombs. The F-16 was a single-seat, single-engine jet, making the LANTIRN mission particularly challenging. I remember one night in Korea, I was on a checkride, or evaluation of my ability as a flight lead to get our simulated bombs on target. It was night, there was red air (simulated enemy fighter jets), and I was certain that I was as busy as a human could be—leading four F-16s, with enemy fighters trying to shoot us down, trying to hit our target with manually aimed laser-guided bombs, at low altitude and high speed, all while trying to avoid the one real-world threat that had a 100 percent chance of killing us if we messed up: the ground. I didn't see how a human brain could process more information, more quickly, than what was called for on this particular night flight.

That is, until I got to NASA and flew space shuttle launch simulations, keeping track of main engines and comm and electrical and hydraulic and propellant systems—all while communicating with my crewmates, as well as mission control. I had found something busier than being an F-16 flight lead. A really good crew and flight control team working together was, in my humble opinion, the pinnacle of what humans are capable of. It was a thing of beauty that will never be matched again, because the new capsules coming on line are much simpler and more automated that the shuttle was.

I made my fair share of mistakes during those shuttle years. One particularly memorable blooper occurred at NASA's Ames Research Center in Mountain View, California. To get there we would fly our T-38 from Houston to California, stopping for gas twice. Ames had a full-motion simulator called the VMS that allowed us to practice shuttle landings from approach to touchdown, all the way to wheel stop. The unique feature of this sim was that it moved up/down/left/right more than sixty feet, giving the crew realistic motion cues of landing and rollout. On this particular occasion, I was late, and immediately after landing my T-38, I jumped in a van and sped to the VMS. As the shuttle pilot, I was responsible for deploying the landing gear and parachute, while my classmate, the commander, actually landed the shuttle. So I jumped in the simulator breathless and we began the first run, diving toward the runway at a 20-degree angle. As my colleague pulled the orbiter's nose up for a gentle landing, my job was simple—arm the landing gear as we descended through 2,000 feet above the runway, and deploy them at 300 feet. After touchdown I would deploy the drag chute to slow us down.

Unfortunately, I was too rushed and hadn't let my brain slow down and catch up to what we were doing. As we dove at the runway, I made the standard calls to the commander—"10,000 feet, 7,000, 5,000, radar altimeter good, 3,000, 2,000 feet"—at which time the commander called out, "Arm the gear." I promptly pushed the buttons to deploy the drag chute instead of arming the gear. Well, even if you're not a rocket scientist, you can probably imagine that the shuttle doesn't fly too well with a giant parachute trailing it, and

Robotic training in the Cupola simulator, Building 16, Johnson Space Center, Houston, Texas, USA, Earth.

despite my best effort to jettison it immediately after uttering "#!@$&!$," we promptly crashed.

You know what? I never did that again. For years after that incident, I never ever made a mistake with the landing gear, or the parachute buttons, or the entire landing sequence. So, despite the fact that it was funny and embarrassing and cost me beers at dinner, that incident served a useful purpose. Making a mistake in the simulator was a *really* good thing—if the crewmember realized it and internalized it. The best way for me to learn was by doing and making mistakes. If I went through a sim and everything went great, I didn't learn anything profound, but debriefing mistakes hammered home the most important lessons. You just didn't want to make a habit of doing boneheaded things like crashing the shuttle, or the boss would give you some career counseling: Find another career.

I was an evaluator pilot in my F-16 days, giving my colleagues checkrides, ensuring they could fly safely and get the mission done. During one particular emergency procedure evaluation, or EPE, I ran one of my squadron mates through a simulated engine failure where he had to land his single-engine Viper with no power. He didn't do a great job, and I dinged him in the debrief about his flame-out landing technique. Well, a few weeks later the two of us found ourselves flying over northern Iraq when he had an actual engine failure, becoming a 30,000-pound glider over enemy territory (Saddam Hussein was still in charge when this incident occurred). And he did an absolutely amazing job, nursing his sick jet back to a remote airfield across the border in Turkey with a razor-thin margin of error, earning himself a well-deserved medal and even national press a few years later when the incident became unclassified. That was a perfect example of why we train in the simulator—you make your mistakes and get familiar with complicated procedures so that when they happen for real, and your butt is on the line and the stress level is maxed out and you're sucking seat cushion, you do what needs to be done.

The same was true in the space shuttle. Some used to say, "Train like you fight." I always said, "Train much harder than you fight." Then when launch day comes, you'll be ready for anything.

Plus, there's no Sim Sup on the real space shuttle. Thank God.

CHEZ TERRY

Styling the Hair of a Superstar Crewmate

When I signed up to be an F-16 pilot, there were many things I expected to do as part of the job description. Learning to fly. Survival training. Studying enemy fighter jets. Moving around the world. Putting my life at risk. These are just a few of the expected duties of an Air Force pilot. However, of the thousands of skills I've had to learn over the years, cutting my (female) crewmate's hair was unique, something I never thought I'd do.

For those of you who are American, I challenge you to name any astronauts not named Neil Armstrong, Buzz Aldrin, or John Glenn. . . . Well? Anything?

That's what I thought. There are a lot of us, and we just aren't that famous anymore. Which is a good thing; it means that spaceflight has become routine. But for other countries, it is a different story, and when I was assigned to fly with Samantha Cristoforetti, I learned just how big that difference was. You see, in addition to being amazingly skilled at foreign languages, she was the first-ever Italian female astronaut. On top of that, she has an engaging personality, and her first spaceflight completely captivated the nation of Italy. After getting back to Earth our crew went on a postflight visit to Italy, and when our train would pull up to the next destination there was always a crowd of people waiting to cheer for her. When I give talks, I always ask the audience if there are any Italians present, and if so, "Do you know Samantha?" The answer is inevitably a resounding "*Si!*" She is a big deal in *Italia*.

In that context, it was going to be important to take care of her hair while in space. Sam is in no way overly concerned about such things under normal circumstances; she was a pilot in the Italian Air Force and is very laid back, but she was going to be a rock star during our half year in space, constantly on Italian television and doing media appearances, and would need to keep up her hairstyle. Keeping my hair looking acceptable was pretty easy. I put the electric trimmer (which came with a built-in vacuum) on number 2, ran it over my noggin, and thirty seconds later voilà, hairstyle complete. But for Samantha, it would be a whole different ball game. Lots of women have flown in space since the 1960s, but to the best of my knowledge, they have always just let their hair grow or kept it in a bun, waiting to get back to Earth to visit a proper hairstylist. But not us—we'd be blazing new trails during Expedition 43.

As for me, my palms were sweaty. My heart was racing. This was going to be unlike anything I had ever done in space.

Before launch, Samantha told me that I'd have to visit her hairdresser in Houston and get trained before I'd be qualified for launch. I was game, and our training manager put it on our schedule. One day after work we went to the salon and I met her hairdresser, a lovely Vietnamese lady named Linh, who was quite experienced. And patient. As Samantha sat in the chair, Linh explained the myriad equipment and procedures required to do a proper lady's hair styling. First there are the clips that you need to put in the hair. Next is the proper grip of the scissors—with thumb and ring finger. Who knew? It's important to use a comb and not a brush. Keeping the hair damp is another key. You start cutting from the center out, holding the hair between your fingers, with the help of the clips. Samantha liked her hair short and layered, so I had to cut it angled at the end, not in a straight line. You have to keep track of which batch of hair has already been cut. Frequent stop-and-assess breaks are required. The hair over the ears had to be trimmed at a sharp angle, à la *Star Trek* style. And the list goes on and on.

Samantha and her hair stylist had the patience of Job. It took me two and a half hours that afternoon to learn these skills. I was so deliberate about each and every cut. It was unlike anything I had ever done in my life before, and I just didn't want to mess that up.

Fast-forward to the midpoint of our mission, and her hair was getting a little long. It was time. I gathered the hairstyling equipment, got the vacuum ready, cleared out an area in Node 3, set up the video cameras, and gathered

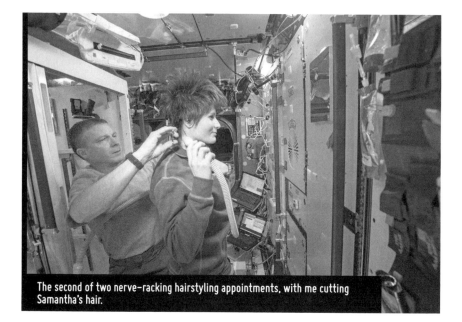

The second of two nerve-racking hairstyling appointments, with me cutting Samantha's hair.

the crew. Anton would be the vacuum operator—sucking up pieces of hair as they shot off after I cut them. Samantha would merely have to float there and smile. As for me, my palms were sweaty. My heart was racing. This was going to be unlike anything I had ever done in space.

The actual operation went smoothly. I put the clips in, and Anton kept her hair wet with a spray bottle as I floated back and forth, working on my masterpiece. I found the task much easier in space because her hair was standing up, so it was more forgiving to sculpt her hair in layers. The whole process took only about thirty minutes, which was much quicker than in Houston, and we had a lot of good laughs. We ended up doing this twice during our 200-day mission, and both times Samantha seemed to be happy with the results. Whew!

I've been very fortunate in my career. I've done some pretty crazy and hair-raising things. Flown F-16s over Iraq and Korea. Piloted a space shuttle. Flown a rendezvous mission. Done spacewalks. But in all seriousness, by far, without a doubt, the most stressful thing I've ever done in my life was cut Samantha's hair. Because if I'd screwed that up there would have been millions of Italians angry at me. And that's probably not survivable.

IT'S NOT ROCKET SURGERY

Medical Training for a Spaceflight

Several decades into my career as a fighter pilot and astronaut, I finally had a chance to do something I had always wanted to do. In the early 1990s, when the Soviet Union collapsed and peace broke out, the Air Force offered a very generous "early out" program, where members could simply leave the service even if they had years left on their commitment. Because there was no evil empire to fight anymore, we didn't need as many people. Several of my Air Force Academy classmates and F-16 buddies took their offer. One went to play professional baseball and made it as high as Triple-A in the minor leagues. Another pursued a career as an actor and was in a Bud Light Super Bowl commercial, which impressed us all. As for me, I went and talked to the squadron flight surgeon about going to medical school, though I decided to stay in the Air Force and continue flying. I guess it worked out in the end.

When I was assigned to fly as a shuttle pilot on STS-130, one of my duties was to be the CMO (Crew Medical Officer). For me this was very cool. Many of my astronaut colleagues were not that excited about the idea of being CMO, and on STS-130 there were no doctors, only pilots and engineers, so I felt lucky to have this duty.

NASA trains nondoctor astronauts for some of the skills of an EMT (emergency medical technician). We can do IVs, shots, and medical exams, perform CPR, etc. We are not given the full training that a nurse or professional EMT would get, but we are able to take care of the basics. Most important, if there were a medical emergency we would be able to communicate with the flight surgeon on duty in mission control, being his or her eyes, ears, and hands for the space-bound patient.

All of my training as space shuttle CMO occurred at the Johnson Space Center in Houston, using medical dummies as well as human volunteers. Yes, they asked engineers to volunteer and let us astronauts practice giving them shots or starting IVs. I'll never forget one volunteer who came in to get his blood drawn by me. I was wearing a white lab coat and had a stethoscope around my neck and looked like George Clooney in *ER*. I was handling everything quickly and professionally and must have seemed like I knew what I was doing. Just before I started the IV, he nervously asked, "You're a doctor, right?" I replied, "No, in fact, I've never done this before." He literally turned white and his mouth gaped open. I went ahead and plunged the needle in, finding the vein on the first try.

I learned an important lesson that day. No matter the situation, act like you've been there before. Never let them see you sweat.

My first shuttle CMO experience came and went quickly, without fanfare. A few years later I was once again assigned as CMO, for my long-duration mission, only this time I had a real treat. I got to spend a week at several hospitals in downtown Houston, doing FMT (Field Medical Training). I worked in an emergency room, an operating room, an ICU (intensive care unit), and even an eye clinic. And I loved every minute. I was so pumped about being a doctor that week that I actually went to Barnes & Noble and got a MCAT (Medical College Admission Test) book. More on that shortly.

That week in the hospital was a highlight of my career. I had the honor of working with a human cadaver, ridding me of any medical inhibitions. In the emergency room I got to assist in some very serious situations, including bandaging severe burn victims, stitching up some gaping wounds, observing chest tube insertion (I had no idea how much water could squirt out of a human chest), holding a dear elderly lady's hand as the orthopedic surgeon set her badly broken femur after a car crash, and performing ultrasound exams to quickly establish a new patient's condition upon arrival. I felt like I'd seen most major ER situations except delivering a baby (and I wanted to do that but my trainer told me I wouldn't be doing that in space so there was no need).

The inpatient side of the hospital was equally fascinating. I helped with intubation, catheterization (not my favorite), and IV placement and did rounds

"Dr." Virts, Crew Medical Officer. My patient didn't fare too well.

with our instructor-physician. He was eager to teach, and I was eager to learn. Something that was both fascinating and sad were patients who had wounds that allowed us to see inside their bodies. I'll never forget two in particular; one was a gunshot-wound victim and one had been riding a motorcycle, and both had significant internal areas of their bodies exposed. I learned a lot about what the inside of a body looks like thanks to that invaluable training, thanks to those poor patients. If I ever had to deal with an actual emergency, having had that experience would have made me infinitely more capable.

I also came home from that experience and gave my kids four brief words of advice: 1) Don't ride motorcycles. 2) Don't drive like an idiot.

3) Don't be around guns. 4) Don't smoke. If you avoid those activities, your likelihood of ending up in the ICU or at the morgue prematurely are seriously reduced. My time in the ICU taught me what COPD (chronic obstructive pulmonary disease) is and the grim future that many smokers have to look forward to. I'm sure my teenagers didn't believe me, because I'm dad and I don't know anything, but if you heed this advice you will have a great chance of avoiding some awful experiences.

These skills were useful in space, even without medical emergencies. As part of our normal activities I had to give shots, draw blood, take urine and saliva samples, etc. The most common medical tasks were performing various scans. I did a monthly set of ultrasound measurements on my brain, my heart, and even my eyeballs to determine how eyesight was affected by fluid shifting due to weightlessness. I would strap myself down to a large storage bag with a bungee cord to stabilize myself, set up a video camera for the medical team down on Earth to observe me, and operate the ultrasound device while searching for brain activity. I wish I were kidding, but I'm not—it was very difficult to ever find any brain activity behind my thick skull. Imaging the shape of my cornea and determining blood flow in my heart were much easier tasks. I took pride on being able to find the required targets quickly, without too much guidance from the medical team on Earth.

One interesting difference between space and ground ultrasounds was the fluid used. Earthbound medical technicians use a Vaseline-like gel, allowing the ultrasonic signal to be transmitted from your body to the sensor. It always seemed like they kept that gel chilled in the fridge to generate screams from patients when it touched them. In space we have endless containers of that same gel. A few years ago Don Pettit, a colleague of mine and probably the smartest astronaut in an office full of smart people, discovered that water works very well for ultrasounds; it transmits the signal without the mess (or screams) from gel. So that's what I used for my scans—a blob of floating water to help the sensor contact my eyeball or brain or heart.

Another important scan that our whole crew did was the OCT (optical coherence tomography). It's a machine that uses light waves to look into the back of your eye to image your retina. And because some astronauts have had problems with eyesight, we spent a lot of attention trying to determine how our eyeballs were being affected by weightlessness. It was much more of an art than a science trying to use this machine. And it was also a two-person job. The scan-ee would stabilize themselves and press their face up to this machine, placing their eyeball at just the right angle to the sensor. It took a long time to go through the series of scans for each eye—maybe an hour. I actually developed a reputation for falling asleep while getting scanned, and then the machine couldn't see into my eye because it was closed! I recommended they send up some energy drinks for this experiment.

One final eye scan was the fundoscope. This device looks like a blaster from *Star Wars*. You press the pointy end against your eye, and it takes visual and infrared images of the back of the eye (retina). As with the ultrasound, I took pride in being able to get this scan done quickly. Julia Wells was the technician on the ground who would talk me through these scans. She also happened to be responsible for all of my medical training, including FMT. It became a competition to see how quickly I could get an image taken. Hopefully faster than my crewmates could.

Another skill that I trained for was basic dental work. Like medical training, I really enjoyed dental training, which took place at the University of Texas School of Dentistry in Houston. We learned only the basics: how to remove a tooth, replace a filling, etc. No root canals or braces while in space. Later, believe it or not, one of my crewmate's fillings fell out, and I had the opportunity to do the first-ever filling in space! This happened on a weekend, and before I performed the procedure, I called my family dentist on Earth to review the details of this task because I couldn't get hold of the NASA dentists. Performing a tooth filling was definitely something I had not expected to do, but it was fun.

I loved medical training before flight, and I loved being the Crew Medical Officer while in space. Which brings me back to the MCAT story. I went to Barnes & Noble to check it out, and after about ten minutes of skimming through a practice medical school entrance exam, I came to my senses and put it back on the shelf. I'd be keeping my day job.

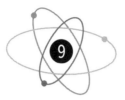

MOUSE MATTERS

Live Animal Experiments in Space

Science is the mission of the International Space Station. It's the reason why a fifteen-nation partnership came together several decades ago and spent many billions of dollars building and operating this magnificent and complicated and dangerous orbiting outpost. During my 200-day mission, we ran more than 250 experiments, spanning every discipline of science—biology, chemistry, physics, astronomy, medicine, engineering, psychology, materials science, combustion science, you name it. Some investigations were very simple, and some were quite involved. Being a pilot, not a trained scientist, I enjoyed my time spent acting as the eyes and ears for scientists back on Earth, performing their experiments, as they nervously watched this fighter pilot operate their precious cargo.

One of the most interesting and potentially impactful experiments we did on the station was something called rodent research. Yes, we had mice on the station. As you can imagine, this led to plenty of jokes. No, we didn't have a cat on the station. No, we didn't feed them cheese from the Moon. No, we didn't have mousetraps in case they were badly behaved. No, they didn't earn *Mousetronaut* wings, though that is the cool title of a bestselling children's book that Mark Kelly, a fellow astronaut, wrote.

Mice are extremely common in medical research on Earth. In fact, rodent research projects commonly order cohorts of more than 1,000 mice at a time from the mouse store. I checked and was unable to find them on Amazon, but maybe someday. As mammals, mice are good analogs for human physiology, including pharmacology; you can test medicine on them to predict its effect on humans. Also, when living creatures go into space, they have a variety of interesting physiological reactions. Some biological processes are accelerated and others are more intense, such as the virulence of pathogens. Others are hindered, such as immune system effectiveness. All of this combines to make space an ideal place for medical research and mice ideal subjects.

Before astronauts are cleared to work with these little furry creatures, we must be trained, and that is a very involved process. The first thing we learn is that the animals must be treated humanely at all times. There are many processes in place to ensure that the animals are well cared for, every step of the way. Another critical aspect of training is learning how to handle the rodents. They are provided with enrichment activities that allow them some diversions during their time in space, including exercise. They are monitored continuously, and any animals deemed distressed are immediately tended to. NASA's mice are the luckiest rodents, on or off Earth. I'm sure my pet dogs wish they were treated as well as the space mice, with the exception of dissections.

After learning the basics of rodent care, it was time for actual medical training. First was how to scruff the animal, i.e., grab that wily rodent while she (almost all of the subjects are female) darts around. On Earth it is tough to catch a mouse because they can run on the ground and they're highly motivated to avoid capture. In space it's even tougher because you're awkwardly reaching into their habitat and you have thick work gloves on, because those little suckers bite. Once you have her, there's a very precise way to hold the animal, grasping her skin on the back of the neck and spine.

After capturing a mouse come several quick steps. First is to move them from their habitat to the work area, known as the glovebox, where we do science activities involving dangerous or messy experiments. Some of the mice are dissected, some are given X-rays to see how their bone density is changing in weightlessness, and some are returned to Earth. The X-ray machine is interesting—the mice are sedated and then placed on a tray for the X-ray. For the dissection they are euthanized via injection. Once the euthanization has been verified and double-checked, the dissection begins.

My rodent training was no less interesting than working with a human cadaver at a local hospital about a year before launch had been. It was fascinating to see how an animal body works, in detail. Astronauts who were trained medical doctors were very much used to such activities, but to this fighter pilot it was all new, and I loved it. Some of our specialized medical hardware included a centrifuge that was used to prepare blood for storage, and a freezer called MELFI (Minus Eighty L? Freezer I?) that was actually cooled to –95°C. I'm not sure how they came up with that acronym, or what the L and I stand for, but MNFF would sound weird. The freezer was used to store all kinds of biological samples as they awaited return to Earth—human blood, urine, and saliva, as well as various samples from rodents, or plants, or any other biological experiments that needed to be preserved for return to the planet.

One of the most involved parts of the rodent experiments was filming them. We filmed literally every second of everything we did with mice, usually with two different HD (high-definition) video cameras running simultaneously, generating many gigabytes of video that would take several days to downlink to the scientists.

Surprisingly, the most challenging part of the mice experiments was finding, retrieving, and stowing all of the equipment required. Rodent research required several enormous MO-1s (NASA acronym for big-ass bag) full of gear, each one bigger than a refrigerator. It would take us hours to deal with the logistics of setting up and tearing down rodent research, and this was the area where I had the most suggestions for improvement during my debriefs.

My big hope for our rodent research program is that it bears fruit for the government and pharmaceutical researchers who work on it. There is real promise for improved care for humans with bone and muscle disease, speedier recovery for those with traumatic wounds, as well as other possible drug breakthroughs. We should all be thankful for the role that these animals play in making life better for humans.

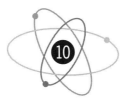

CLOTHES MAKE THE ASTRONAUT

Packing for Six Months in Space

Though here are many things to be done before launch: learning how to pilot the shuttle, how to fly a rendezvous mission, how to maintain the spaceship, how to install and operate the payloads you bring to space, how to perform spacewalks, how to operate the photography and video equipment, etc. One of the most important, however, is picking out what clothes to wear. The good news is there isn't much choice, but nonetheless there are important decisions to be made. On a two-week shuttle flight it wasn't too critical, but on a 200-day, long-duration flight, wearing the wrong thing every day would have gotten old.

Let's start with shirts. On my shuttle flight NASA provided me a new shirt to wear every day, which was kind of ridiculous, but I didn't complain because it was nice to keep them after the mission as a memento. They were Lands' End cotton polos; we couldn't use modern polyester fabric because it isn't fireproof. In space you don't get dirty or sweat much unless you're exercising, so those polos were still in good shape after only a day or two of wear. However, on my long-duration station flight I was provided one polo per month, which I only wore for special occasions, like media events on camera. They gave us basic T-shirts for daily wear, and color selection was simple enough for this fighter pilot—black, green, red, or blue. They were fine, and comfortable, but not very fashionable. I always thought they should at least have put a mission logo on them.

Bottoms were either full-length pants or shorts, tactical style with lots of pockets and Velcro added to them. Honestly, they were genius. I came up with a system of keeping my Leatherman multi-tool in one pocket, fresh CF memory cards for cameras in one pocket, used memory cards in another, Chapstick in one, headlamp in another, etc. Those pockets saved me countless hours of rummaging for stuff, because I always had what I needed with me. On Earth it would have been miserable to have all of that stuff tugging down on your pants, but in space it floated and wasn't an issue. A pair of shorts would last for many weeks, usually months. I rarely wore long pants, saving them for media events.

Exercise clothes were also really important, though we didn't have much choice in that matter either. We got a fresh pair of running shorts every week, along with a new Under Armour shirt every two weeks. The running shorts would get a little stinky, but there was a new pair after a week. On the other hand, the shirt got stinky after one exercise session on the treadmill, and there were still thirteen days to go to the next shirt! Those things got ripe, to say the least. I actually sent a used one back to Earth on a SpaceX Dragon cargo ship, tightly sealed in Ziploc bags covered in duct tape. When I got back to Earth a few months later, I opened it and almost died, it smelled so bad. In fact, one of the experiments I took part in was to test out a new type of odor-free exercise shirt, made from antistink wool fabric. I was skeptical; I thought it would be too itchy, but I gave it a shot anyway. I wore

> Underwear was an interesting subject. A few weeks before my launch, a Cygnus cargo ship had blown up immediately after launch, and with it went half of my clothes and underwear for my imminent half-year mission.

that shirt every day for a month of exercise, two and a half hours drenched in sweat per day, and after thirty days it didn't stink. I was shocked. That wool really worked.

Underwear was an interesting subject. A few weeks before my launch, a Cygnus cargo ship had blown up immediately after launch, and with it went half of my clothes and underwear for my imminent half-year mission. So that was bad. Luckily, a choice I made when picking out my clothes ended up saving me. Crews go to the Flight Crew Equipment office about a year before flight, both in Houston and in Russia (and by that I mean Star City, which is how all US astronauts think of Russia), to select a whole host of personal items for flight. You sit down with experts and pick out clothes, tools, headlamps, notebooks, razors, sunglasses, toothpaste, and deodorant—basically, any personal items. However, you can't just pick anything you want; you need to select your items from a fairly limited, flight-approved menu. Luckily, I had a philosophy of variety; I didn't want to pick 100 percent of one thing and end up in space and discover that it didn't fit or work, so I tended to pick different sizes or types. I chose mostly L shirts, but also a few M and XL. Mostly shorts, but two pairs of long pants. Mostly thick socks, but also a few thin ones. You get the idea.

When it came to underwear, I picked mostly the standard tighty-whitey type, but I also picked some other longer briefs as a backup. Fast-forward to my first week in space, and I was in pain! It felt like something was squeezing my boys, and not in a good way. At first I thought it was my body adapting to zero g, but then I realized it was my underwear. So I switched to the briefs, and they were great. It turns out that underwear can last a really long time in space, and I went 200 days using one pair of my alternate briefs every two weeks. I ended up with a million extra tighty-whities, so I used them as towels and spare rags. It was a little awkward, but I was in space, and you have to be resourceful and use what you have. The *Apollo 13* guys would have been proud. My philosophy of underwear variety ended up giving me the margin to survive the Cygnus explosion.

Another important piece of clothing was running shoes. Exercise was such a critical part of daily life on the station that these were extremely important, and shoes were one of the only things where we had freedom to get what we wanted. I took two pairs of the same brand of running shoes that I used on Earth. I didn't wear shoes most of the time, but I did for the treadmill and weight-lifting machine (for the bike I used standard cycling clip-on shoes). After I got back to Earth, I sent those running shoes off to a prominent athletic gear company for them to analyze how they'd held up in space. They put them through a vigorous X-ray and MRI testing protocol, and their engineer told me he's never seen a piece of athletic gear revered in their lab as much as my space running shoes were. It turned out that they were almost like new, which surprised me because I did so much running. Maybe it was because running on the treadmill in space had a lower load factor than on Earth; we were held to the treadmill with a bungee cord that pulled me down with the equivalent force of less than 70 percent of my body weight. Or maybe it was because the treadmill itself was floating, not fixed to the wall, giving the shoes more cushion than they'd have on Earth. Whatever the reason, those shoes could have lasted indefinitely. It's one of the few areas where space is cheaper than Earth!

Getting some personal items and clothes back to Earth was not a problem on my shuttle flight, but on the Soyuz there was very little room for returning items to Earth. The good news was that the equipment returning on the SpaceX Dragon needed packing material, so we used worn clothes for that purpose, and sometimes we even got some of those items back. I ended up getting a few exercise shirts and polos. When I die my kids will get a chest full of old space stuff, and they will be able to know what Dad smelled like while running on the ISS treadmill, all those years ago. I wonder if they'll realize just how important it was to have two types of underwear?

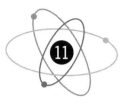

ASTRONAUT CROSSFIT

Physical Training for Spaceflight

Let's face it: Going to the gym ain't easy. For me the biggest impediment is time—it takes me a few hours to get dressed, drive to the gym, do some weight lifting and cardio while watching Netflix, drive home, and shower. It's not a twenty-minute activity. Even going for a run in the neighborhood ends up taking an hour by the time I stop sweating in the Houston humidity and shower. So, is it worth it? And how do busy astronauts fit gym time into their schedules?

The most important thing for me was to realize that fitness was critical. In space, your bones and muscles don't have to constantly fight against gravity like they do on Earth, and that leads to atrophy. You may be lounging by the beach as you read this, and as you do you are getting a workout simply by holding up your head and arms, however light they may be. That 24/7 effort doesn't exist in orbit, with serious consequences. During the Russian Mir space station program in the 1980s and '90s, researchers learned that cosmonauts' bones would atrophy about 1.5 percent per month, no matter how long they were in space. It was as if they were on a straight-line trajectory to becoming jellyfish. In addition to losing bone density, the space travelers also lost muscle mass, especially in the lower body, which led to serious problems upon return to our planet. I know two people who have broken their hips who weren't over the age of ninety, and both had recently returned from a long-duration spaceflight.

To combat these bone and muscle problems on the ISS, we came up with a very effective exercise protocol. Like any good Earthbound workout regime, it involves two basic components, aerobic and anaerobic. The aerobic component was fulfilled by TVIS (NASA acronym for treadmill) and CEVIS (NASA acronym for bike), and the anaerobic was fulfilled by ARED (NASA acronym for weight-lifting machine). More on those later, because before you get to space you need to get in shape on Earth.

One of the benefits of being an astronaut is having access to an amazing gym. When I first showed up at the Johnson Space Center in 2000, it was a decaying relic from the 1960s, but the newly built astronaut gym is a palace that would make any professional football team jealous. It is roughly the size of a grocery store, and there are rarely more than a handful of people working out. Every imaginable aerobic and free weight machine is available. The most interesting machine there is called the zero-g treadmill, in which you wrap a giant bubble skirt around your waist and inflate it with air. This lifts you off the treadmill while running, which means less leg impact, making it possible to run with less wear and tear on old knees. There is a treatment room where our flight surgeons could come and perform exams, a lap pool, and hot and cold tubs. There is also a room full of mock-ups of the CEVIS, TVIS, and ARED space hardware for us to train on. Overall, it's a pretty impressive facility.

More important than the hardware were our ASCRs (pronounced "ace-ers"), the NASA acronym for personal trainers. These guys and gals have backgrounds as top collegiate strength and conditioning coaches, and their job is to get us ready for spaceflight. Once we get to space, they tailor our daily workout program and track our progress as we log our exercise sessions, sending us updated plans. They also watch our form and technique via downlinked video from the ISS and critique us, making sure we're squatting deep enough, running on the treadmill with good form, etc. Station crews have a short tag-up with them about once per month, something that I always enjoyed, talking not only about exercise techniques but also about the latest sports scores on Earth, family updates, or other gossip. Those calls were a nice break from the monotony of repairing hardware and performing experiments on the ISS.

Besides basic bone and muscle health, there was another important reason to get and stay in shape: spacewalking. Putting on that beast of a spacesuit and moving around in it for eight to nine hours at a time is a serious workout, using a significant amount of upper body strength, endurance, and hand strength. That strength doesn't just happen on its own; it requires a lot of time at the gym, especially doing exercises that focus on your forearms and hands, like pull-ups, kettle bells, etc.

With that as background, what exactly was our workout plan to prepare for spaceflight? Every astronaut is different, from thirty-something marathoners and mountain climbers to sixty-something "experienced" astronauts, and the average age of crewmembers flying in space is probably around fifty. The vacuum of space is not politically correct. It doesn't care what your age or gender is. And the inevitable toll weightlessness takes on your body doesn't hold back for those who don't have time to work out. Everyone understands this, and most make a concerted effort to get in shape preflight and stay in shape once in space, and to rehab once back on Earth.

One of the best parts of working with an ASCR was having a personalized workout, as well as motivation to keep going. There is always a WOD (workout of the day) posted on the gym bulletin board that is a quick and intense body-weight program: pushups, pull-ups, lunges, sprints, stretches, etc. If you are fast you could finish in twenty to thirty minutes, but it would wipe you out. There is also a more formal weight-lifting program, focused on Olympic-style lifts. There are also plenty of aerobic opportunities, with miles of outdoor running trails, indoor machines, and a pool.

In addition to daily exercise, astronauts are evaluated with an annual fitness assessment. It includes everything from a 1.5-mile timed run or 800-meter swim to max-effort bench and leg presses, pushups, pull-ups, crunches, hand-strength and flexibility measurements, and everyone's favorite, a shuttle run around a series of pylons.

A critical part of our fitness program is rehabilitating once back on Earth. After landing in Kazakhstan, there were folks waiting for me next to the Soyuz to begin the torture (I mean rehab) program within minutes of landing. Following a twenty-four-hour, three-leg journey back to Houston, the very first thing I did was go to the astronaut gym for a grueling workout. It included trying to make my way around those pylons while being completely dizzy and standing on wobbly gel pads that forced me to seriously concentrate on balance, while my ASCR stood ten feet away, throwing a medicine ball at me. That's hard enough to do under normal circumstances, much less after spending half a year floating in weightlessness. But those exercises, focusing on balance, coordination, and core strength, expedited my return to Earth form.

I made it a point to visit the gym every day for several months after getting back to Earth, and it seemed to make a difference. I didn't suffer the serious back pain that some of my colleagues had and I never had a serious orthopedic injury, even though Achilles tendon tears or muscle problems have been common injuries. Within a week of landing I was doing twenty pull-ups, and most of my weight-lifting strength was at 90 percent of my preflight numbers. Although I was at the gym every day, I didn't push it; I knew too many colleagues who had done so and ended up with a torn muscle or ripped tendon. I resolved to work my body, but not push it too far. Better to work out every day at a decent level of effort than to push it to the max, get hurt, and end up missing a month of exercise while I healed. This technique worked well for me, and I was amazed how quickly I regained my strength with no injuries. Some of my friends who weren't as diligent at the gym ended up having a much rougher time.

The adverse effects of weightlessness on our bones are similar to those of osteoporosis, the disease that older women (and occasionally men) are at risk for, so whenever I do a talk for a large audience I promote the benefits of resistive exercise, even using small five-pound dumbbells, to keep bones healthy, especially for women. Waiting until old age, when you are hunched over from that terrible disease, is too late, and starting a habit of both aerobic and anaerobic exercise is good for everyone at any age. And though I was lucky enough to have the astronaut gym to prepare me for spaceflight, I recommend that you find a place near you to work out. You'll be ready for spaceflight if need be, *plus* you'll look great at the beach for spring break!

JET LAG
(AND SPACE LAG)

Adapting Your Circadian Rhythm

There are a lot of skills that you need to learn when you become an astronaut. Most are what you'd expect—flying, spacewalking, doing experiments, repairing equipment, etc. Others are surprising but make sense: learning Russian, training to perform medicine, learning how to make an IMAX movie. A few are downright unexpected. One of those skills is how to deal with time zone changes, both here on Earth and in space. There were so many instances when my body's internal time clock felt like a tennis ball during a match, getting knocked back and forth, usually suddenly and painfully.

I had traveled overseas ever since I was a high school exchange student in Finland, spending time in France during college and then being stationed all over the world while in the Air Force. I thought I knew the basics of dealing with changing time zones. It messes with your body clock for a few days and then you're adjusted to the new time zone. Well, once I was assigned to my long-duration mission I learned a whole new meaning of jet lag. I was constantly flying to Europe, Japan, Canada, and most of all Russia for required training. These trips were often combined; for example, I might take a Europe-Russia trip over a four-week period. A few times I even did a "slow orbit," flying to Russia, then Japan, and then back to Houston, undergoing twenty-four hours of sleep-shifting in one month-long, round-the-world business trip.

This circadian ping-pong went on for about three and a half years, and it really took a toll on my body. NASA flight doctors gave us briefings about how to best manage our body clocks. For me, the most important factor was trip duration. If I went to Russia for three or four weeks of training, I would completely shift to their time zone. If it was just a one- or two-day quick trip, I wouldn't bother; I'd simply sleep when I felt tired. There were several tricks to completely adjust to the new time zone. I thought going west was always easier than going east, because it's basically like sleeping in late, or getting a few extra hours in the day, which is never a problem for me. Flying to the east was always harder because it was like losing those hours, or waking up early, which I'm not a big fan of. Of course, if you're going to Japan or China, with a twelve-hour shift, it doesn't matter if you fly west or east—you're hosed either way!

I learned a motto while at NASA—"better living through chemistry," an inside joke about the importance of taking the right medication at the right time. For example, in our Vomit Comet, the zero-g training aircraft, it was very important to take the anti-motion-sickness medication. Likewise, when you're shifting time zones there are some medications you can take to help you sleep. Sonata is a relatively gentle sleep medication that helps you get to sleep but doesn't keep you asleep. Ambien is the most powerful sleep med, and I needed only one small pill, sometimes just half a pill, to put me out for hours. One time I really needed to fall asleep quickly because I was going to be waking up at midnight to work at mission control, so I took two Ambiens. Big mistake. Yes, I fell asleep quickly, moving from feeling dazed and out-of-body to falling asleep in minutes, and then having a lot of bad dreams. When the alarm went off I felt absolutely terrible, and I had an awful headache, like a bad hangover that lasted twenty-four hours. No more double-dose for me; in fact, I try to avoid Ambien altogether unless I absolutely need to get myself to sleep when my body thinks it's wake time.

Another medication was melatonin, a natural chemical in our bodies that regulates our daily circadian rhythm. Taking melatonin near bedtime wouldn't make you fall sleep, but it would let your body know it was night-time; i.e., it didn't help with the sleep but it did help with the shift. On the other side of this equation is waking up. The military uses several types of strong "go pills," designed to give you energy and wake you up, and Provigil is the one that NASA uses. It is the nuclear option of staying awake, the yang to Ambien's yin.

Before my first flight, NASA gave me a bag of medicine to test at home in order to make sure I didn't have any adverse reactions to some possible medications that might be needed in-flight: sleep meds, wake meds, anti-nausea meds, digestion meds, etc. I went to the pharmacy to pick up my test bag of pills early that afternoon. The pharmacist was handing me the goods, including a little brown vial with just one Provigil pill in it. I told her, "Don't bother with the vial, I'll just take it now." She looked at me as though I had two heads and pushed, "Are you sure you want to take that now? It's after 1 p.m.!" I smugly replied, "I'll be fine, these things never affect me." Well, fast-forward thirteen hours and I was lying in my bed at home on my back. Eyes wide open. Thumbs twiddling. Thinking to myself, "Well, now I know if I need to be wide awake for an entire day, take Provigil!" The lady at the pharmacy was right; it's a bad idea to take one of those pills in the afternoon. Luckily, that was the one and only time I took that medication.

Having learned that Provigil was so extreme, I wanted a less severe measure to help shift my body clock for my first launch, at 0400, which meant going to bed at 10 a.m. the prior morning and waking up at 6 p.m. Because we were shifting our body clocks by twelve hours, NASA had us begin the process of waking up a little earlier each day about a week before the mission. To help with sleep, we moved into crew quarters at the Johnson Space Center, a modern facility where we could sleep, eat, and have offices for training. There were about a million super-bright fluorescent lights in that building, so when it was dark outside our bodies were tricked into thinking that it was daytime. And vice versa; when it was bright outside we sealed off all outside light to make us think it was nighttime. Interestingly, in the Russian winter, there's not enough ambient sunlight because it's so dark, so our flight docs had a special portable blue light in our rooms to make our brains think that it was time to be awake.

This process worked well, but I was still groggy when my body thought it was time to sleep, so I used energy supplements. I was skeptical at first, but my flight surgeon recommended them because they help you feel awake while not being nearly as drastic as Provigil. The ones I used were basically vitamin B shots, with minimum caffeine, and they worked for me. I felt awake without any adverse side effects. In fact, they worked so well that I continued to use them for my next mission; they were particularly helpful in the first few days after arriving in Russia or Japan or Europe, because a five-hour energy shot would wake me up just enough to get over a hump in my circadian rhythm. I even used them before spacewalks or other significant events in orbit that I needed to be wide awake for.

There is another important piece in the sleep-shifting puzzle. For international travel, a business-class plane ticket would get the trip started on the right foot, because it was possible to sleep while lying flat. I just can't sleep well while sitting upright in a normal airline seat. Add in the manspread from the offensive lineman next to you in seat 34F, plus having folks crawl over you to go to the bathroom for twelve hours, and you get the picture—a long flight with no sleep. My ability to stay awake during critical training events a day or two after arriving in Russia/Europe/Japan was directly affected by the quantity and quality of sleep I'd gotten in the preceding days. Getting six hours of lying-flat sleep on those flights made a huge difference in my mental capacity in the following days. I still crashed in the afternoon for several days after arriving, but it was much better after a good night's sleep on the flight over.

Those are my secrets to dealing with jet lag and large sleep shifts. Get a decent quantity of sleep whenever you can, preferably lying flat. Meds like Ambien help get you to sleep when required, though they should be used sparingly. Energy supplements help you wake up, if only a little, when Diet Coke can't do the trick. Seeing bright, blue light when it's wake time, and no light when it's sleep time, is very important to keeping that internal body clock on schedule.

LAUNCH

DRESSING FOR SUCCESS (AND LAUNCH)

A Very Complicated Spacesuit

henever you see astronauts on the news, they are in some kind of spacesuit. During the shuttle days, they were in the orange "pumpkin suits," as we called them. When launching on a Russian Soyuz, they're wearing the off-white Sokol suit. And then of course there's the giant white spacesuit for going outside on a spacewalk. But how exactly do we put those bulky things on, how do they work, and what do we do if there's an emergency?

The process for getting suited up is quite involved; it's not like in most Hollywood movies, where it only takes a minute. Or like in *Iron Man*, where you press a button and a suit magically forms itself around your body. (Someone needs to invent that.) Before my first launch, we got suited up at the Astronaut Crew Quarters facility at the Kennedy Space Center (KSC) in Florida.

The history of launch spacesuits at NASA is both interesting and tragic. Astronauts wore pressure suits for the first few shuttle flights, but they soon transitioned to simply wearing a flight suit and a motorcycle helmet for launch, as though they were flying jets. After the *Challenger* accident, NASA realized that some launch and landing accidents could be survivable as long as the crew had pressure suits that would keep them alive in the event of a massive air leak. So astronauts went back to wearing pressure suits, modeled after the U-2 spy plane suit. The first version was called the LES, and the next was the ACES (NASA acronyms for spacesuit and advanced spacesuit).

The first step to suiting up in an ACES was to put on a MAG (NASA acronym for diaper). Yup, a diaper. More on that later. Next, it was time to put on some long underwear, which protected your body from the rough surface of the spacesuit. Then there was a special blue undergarment with a series of plastic tubes woven into it, designed to carry cool water all across your body to prevent overheating. Finally, it was time to get into the suit, stepping into it from behind. There was a giant zipper covered with a big rubber gasket that needed to be zipped, from your neck all the way down your spine to your crotch. This was very hard to do by yourself, and it was nice to have someone help. There was a skull cap and separate headset, or "comm cap," that were basically the same technology from Apollo, with earcups as well as two microphones. I always used the thinnest skull caps available to give a little more room in the helmet, because my head is so giant. The comm cap connected to the suit via a pigtail cord, and then the suit itself was plugged into the shuttle's communication system. Next was putting on the helmet, which needed to seal properly to the suit to keep us airtight in the event of a cabin leak. The helmet had a clear visor that you could move up and down by grabbing the bailer bar, which was locked down with a little tab on the outside of the helmet in front of your chin. There was also a dark visor that could be raised or lowered in case of bright sunlight.

The inside of the suit had a neck ring, designed to keep most of the air pressure down in the lower part of the suit. It did this by squeezing your neck just shy of the point of suffocation. There were two tabs that pulled that rubber neck ring off your throat, and they were velcroed to the outside of the suit. This made it much more comfortable, but in order to close and lock the visor you had to release the tabs; otherwise the visor would close on top of two cords, which would lead to a bad seal and air leak.

The last step was putting on the gloves, which sealed to the suit using a metal ring and could be removed with the help of a lever to pry them off the suit. ACES gloves are much less bulky and offer a lot more dexterity than the EMU (Extravehicular Mobility Unit) spacewalking suit, but they're also designed to inflate only for emergencies. The suit also had a big strap that went from the front of the neck ring down to your crotch, which pulled your neck toward your knees in a constant partial-sit-up position. Under normal circumstances it wasn't really necessary, but if there was an air leak and

Shuttle Extravehicular Mobility Unit (EMU)

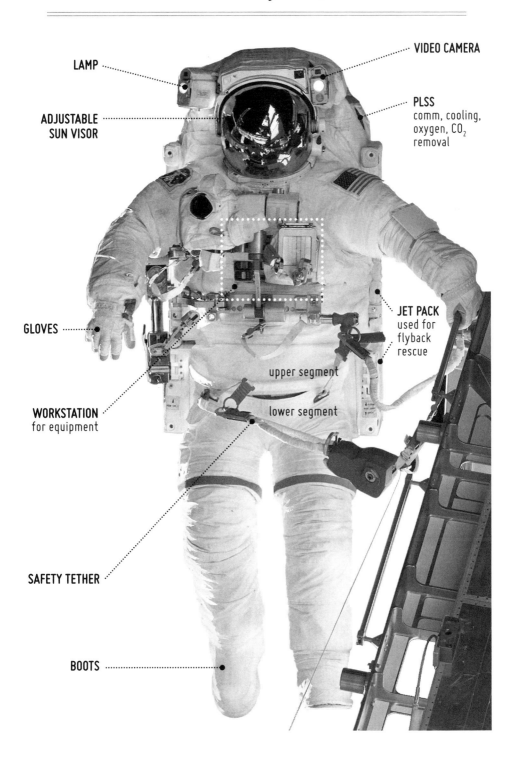

LAMP

ADJUSTABLE
SUN VISOR

VIDEO CAMERA

PLSS
comm, cooling,
oxygen, CO_2
removal

GLOVES

JET PACK
used for
flyback
rescue

WORKSTATION
for equipment

upper segment

lower segment

SAFETY TETHER

BOOTS

depressurization, your suit would try to inflate into a standing-up position. That's not very convenient for a pilot sitting in a cockpit, so that cord would keep us hunched over in a sitting position in the event of an air leak.

Under normal circumstance there was a constant trickle of air into the suit that was vented into the cabin, but you could rotate a knob on your chest to close the vent valve. If the visor was down and locked, the suit would gradually inflate, because that trickle of air would be trapped in the suit, causing you to puff up in a giant orange balloon. We did this to verify the pressure integrity of the suit, though an added benefit of that procedure was that we'd get an idea of what the ACES suit would be like in the event of an air leak.

After suiting up in crew quarters, we walked out to an Airstream trailer affectionately known as the Astrovan, waving to a few hundred NASA employees gathered to wish us well. Then we took a twenty-minute drive to the launch pad, our suits plugged into a cooling unit that circulated cold water through that cooling garment with the plastic tubes. We arrived at the launch pad, took the elevator up to the 195-foot level, made one final bathroom stop in order to avoid using the diaper, walked across the gantry walkway into the white room, put on our parachutes and LPUs (NASA acronym for life preserver), and waited for the closeout crew to give us the thumbs-up to get into *Endeavour*. When it was my time I crawled through the hatch in that bulky suit, disoriented from the shuttle being in a vertical position, 90 degrees different from what I was used to from training in Houston. A fellow astronaut, known as a Caped Crusader or C-squared, was in the cockpit to help us strap in.

The actual strap-in was painful. You had to wedge yourself into the seat, feet on the rudder pedals, trying not to step on or smash any critical switches or the control stick, all while overheating in a forty-pound spacesuit, with a parachute on your back. I had a kneeboard velcroed around my leg that held a few large index cards along with a pencil and Maglite; it was a convenient place to take notes during ascent or once in orbit. I also had a large crew notebook that was full of my personal notes and diagrams collected over years of shuttle training stuffed into a saddle bag full of checklists next to my seat.

The final step was connecting the cooling loop—and ahhhhh, it felt so good. By then I had worked up a sweat, and it was a real relief to have cooling.

The entire strap-in process took an hour. Finally, the C-squared and closeout crew shook our hands, said good luck, and *thunk*, the hatch was closed. The six of us left behind in *Endeavour* counted down the minutes and seconds until we left Earth, each person contemplating the fact that we would be imminently leaving the planet in a trail of flames atop a four-million-pound bomb in the shape of a space shuttle.

One of the more interesting aspects of training in the spacesuit was learning how to perform an emergency egress of the space shuttle. There were eight different modes of egress, including the crew being strapped in on the launch pad all alone, being on the launch pad along with the closeout crew, being in flight after launch, or even gliding back to Earth and having to parachute out of the orbiter. Each of these eight different situations had its own, very specific procedure to safely get out of the orbiter in an emergency. If necessary, the launch director or commander would call, "EGRESS, MODE 1!" or whatever the appropriate mode was for the situation. There were actually five shuttle missions when the engines were shut down seconds before liftoff, putting the crew in a potential emergency egress scenario, though thankfully in each case they were able to get away from the launch pad without incident.

A few weeks before launch, we flew to the Cape to perform what was known as a TCDT (Terminal Countdown Demonstration Test). As part of that training, we practiced strapping in to *Endeavour* while she was on the pad. After the nominal procedures were finished, the launch director called for a Mode 1 egress, which meant we would have to get ourselves out of the orbiter unassisted, then across the gantry and into the baskets. There were a series of four refrigerator-size, wire-mesh baskets at the top of the launch pad, and we crawled into them two by two, visors down and locked and emergency oxygen bottles activated. Once you were in the basket with your buddy, the guy in back would tap the guy in front on the shoulder and he would whack a big guillotine handle, cutting a safety wire that released the basket, allowing it to slide down a wire several hundred feet to the ground, where the basket (with its two nervous astronauts) would safely drag to a halt in a big pit of sand. What could go wrong?

After that, we rushed to a waiting M113 armored personnel carrier (APC), which I was responsible for driving, sticking my head out of the overhead window to see where we were going. There was basically no visibility from inside that ancient army tank, so I had to stand up to drive. At this point our visors were still down to prevent toxic gas from killing us after escaping from a burning shuttle. There was a safe-house underground bunker about a mile away where we could hole up and await further instructions. Either the shuttle would blow up, and we'd (probably) be safe that far away, or NASA would declare that everything was safe, and they'd come get us.

On our TCDT training day, we performed the Mode 1 egress in our pumpkin suits, got in the baskets, and chopped the safety rope, though the baskets were firmly attached and slid only a few inches. Of all the crazy risks NASA was willing to let us take, going for a 300-foot zipline ride wasn't one of them. So we got out of the baskets, went down the elevator, and made our way to the M113. As we got in the old APC, my crewmates snickered at me, because they had all done this training before. I was the only rookie and it was my job to drive the blasted thing, which was done with a handle in each hand, the left controlling the left tracks, the right controlling right tracks. Both handles forward = M113 forward. Left forward and right middle = M113 turn right, and so on. I set out at what I thought was a normal pace, both handles partially forward, and the vehicle lurched ahead, steadily, as everyone smirked. Finally, someone said, "Move aside and let me show you how it's done," and he moved both handles full forward. Boy did we take off, bouncing violently down the road. Kicking up a huge cloud of dust. Running over foliage as if it weren't there. It was a blast, but a driving technique that I wasn't used to, though it was smart—if we were trying to escape a burning four-million-pound bomb, it was best to drive full-speed ahead and tell any policemen who pulled you over, "Sorry, officer, no sir, I didn't know how fast I was going, but OMG there's a space shuttle about to blow up, let me drive as fast as I want!"

Getting to wear spacesuits and strap into space shuttles and drive tanks was awesome. Thankfully, I never needed any of those emergency procedures in real life, but they sure were cool to practice.

WHEN NATURE CALLS

A Spacesuit Has No Fly

So much of being an astronaut is cool. Wearing spacesuits. Flying supersonic jets. Riding rockets. Floating like Superman. Eating astronaut ice cream (OK—that's not really a thing). And when it comes to professions, to quote an awesome AXE commercial, "nothing beats an astronaut." But there are a few things that aren't cool about our job, and one of them is what to do when nature calls during an inopportune moment in the mission.

One of the funniest scenes in *The Right Stuff* is when Alan Shepard was sitting on the launch pad, ready to go, and then launch was delayed. And delayed. And then delayed some more. His mission was only supposed to be a fifteen-minute suborbital flight, so some contingencies weren't planned for. The film depicts this lack of forethought with a beautiful montage of water scenes: a man drinking coffee, water bubbling at the water cooler, folks on the ground going to the restroom, and a clearly agonized Shepard, who had to go. Finally, Gordo Cooper, the CAPCOM, offers relief by radioing to him, "Permission to wet your diaper at any time." Except there was no diaper! Nobody had thought it was necessary for such a short flight.

By the time I flew on the shuttle, we had this issue solved. Yup, astronauts wear diapers for launch, landing, and spacewalks. We are in those spacesuits for such a long period of time that finding a bathroom is impossible and it's not safe to dehydrate yourself, so our flight surgeons tell us to drink plenty of water and wear Depends. And like everything else that we do related to spaceflight, this skill has to be trained.

There are many challenging and difficult things I had to do as an astronaut. Learn Russian. Practice spacewalks. Listen to bad "Houston, we have a problem" jokes for two decades. But one of the more challenging skills was learning to pee in a diaper, lying on my back. Whether you are on the launch pad in a space shuttle or Soyuz, you're on your back for several hours, and that is precisely when you don't want to be stressed out because your bladder is full and you can't go. So early in my training I was given a diaper and told to go home, put it on, lie in the bathtub, and pee. And boy was that difficult—it's just not natural. Your brain is screaming "Hold it!" but you gotta go. What's more, unless that diaper has a good seal, you will have a warm back and neck and hair. It's important to ensure a good fit, to say the least.

We called that big, industrial, designed-to-hold-air-pressure zipper the *jaws of death.* No need for details, but, again, use your imagination.

The only time I actually used a diaper was during my spacewalks. For launches, landings, and countless shuttle and Soyuz training sessions, I was good to go (so to speak). My bladder had enough margin, and diaper operations were not necessary. However, for spacewalks I really wanted to be hydrated because of the physical exertion required, so I drank a lot before going outside. While getting suited up, we go through a prebreathe protocol and use a mask and hose to breathe 100 percent O_2 before finally climbing into the suit. It was while wearing that mask that I took one final trip to the restroom.

A few hours later, I was suited up and waiting in the airlock as the air pressure dropped to zero. It was during that wait, just before going outside, that I inaugurated my diaper. I figured that nature would call at some point during my spacewalk, and better to take care of business during the calm before the storm, chilling out for twenty minutes, waiting for the air to be sucked out of the airlock. This technique worked well for me; I used the diaper at the start of each of my three spacewalks, and never had to use it while actually outside in the middle of a complicated task, which would have required an awkward call to the ground. "Houston, stand by, I'll be off comm for a few minutes; I'll let you know when I'm ready to continue." That was one radio call I was glad I never had to make.

Diapers are not the only interesting bathroom stories surrounding spaceflight. Allegedly, Yuri Gagarin, the first human in space, ordered a pit stop on his way out to the launch pad at the Baikonur Cosmodrome way back on April 12, 1961—and relieved himself on the wheel of his crew van. Because Russians tend to be a superstitious lot, literally every crew launched from Baikonur since then has stopped on the way to the pad to take care of business. As a guy, this wasn't too big of a problem for me, although the Russian Sokol spacesuit zips up from the back, making it not exactly conducive to number one. Use your imagination. But for the ladies, this task was even more difficult. I never asked, nor did I ever understand, nor want to understand, how that worked for them; us guys simply did what we needed to and let our female compatriots do what they needed to. Tradition fulfilled.

The space shuttle launch experience had an interesting twist. The 195-foot level at the launch pad is the floor where the crew walks the plank from the elevator over to the orbiter. The order in which crewmembers went into the shuttle was highly choreographed, and while waiting to board there was a phone to make a last-minute call to your family or order pizza for the launch director. This was a scene repeated many times over the years at the Launch Control Center—knock knock knock, "Domino's Pizza here; we have a delivery for Mike Leinbach." There was also a bathroom, which was a big relief (pun intended). This was significantly easier for male astronauts, with one significant catch. The zipper was in the front of the shuttle suit, as opposed to the back of the Russian suit. We called that big, industrial, designed-to-hold-air-pressure zipper the *jaws of death*. No need for details, but, again, use your imagination. You didn't want to lose a grip on that zipper and let it accidently come crashing closed while you were in the middle of taking a leak. That would have a giant and irreversible impact on potential future offspring.

Being a practical joker, when one of my crewmates went into that launchpad restroom before boarding *Endeavour*, I gave him a few moments to get the jaws of death opened, and then I loudly banged on the door. I think my poor buddy had a heart attack, but it sure did make me laugh, and him, too, eventually, a few weeks later after we were back on Earth. I'm quite certain that I will be repaid at some point, when I'm least expecting it. But that was a lighthearted moment to take the edge off the incredibly dangerous thing we were about to do—ride a four-million-pound spaceship into space.

THE RED BUTTON

How and Why a Shuttle Could Be Intentionally Destroyed

When I was an ASCAN, we took a trip to Florida to see our future launch pads. As part of that trip we visited the Cape Canaveral Air Force Station, a facility adjacent to the Kennedy Space Center, America's primary East Coast launch site, complementing Vandenberg Air Force Base on the West Coast. After the trauma of the *Challenger* accident in 1986, the Department of Defense (DoD) pulled out of the business of using space shuttles for military missions, and the shuttle was henceforth based exclusively at KSC, where there was quite a bit of military/civilian collaboration for launch: weather planning, rescue operations, logistics, and one very important, if obscure, detail: the Red Button. More on that shortly.

That trip to KSC in October 2000 was one of the highlights of my time at NASA. I was beside myself with excitement. We had seen the STS-92 shuttle crew preparing to launch. We stayed at a hotel where astronauts and their families often stayed. I jogged up and down the sand and on the A1A road through Cocoa Beach. We shopped at Ron Jon Surf Shop, buying cheesy T-shirts and stickers. Life was good.

My astronaut class was called "the Bugs." Back in 2000 there was a glut of new astronauts because NASA had hired too many from 1995 to 1998. Space station assembly missions were being delayed, and it was obvious that it was going to be a long time before we could fly. I innocently thought that it might take *five* whole years, so I steeled myself for a long wait. It eventually took our group of astronauts between eight and twelve years, but it was worth the wait. With all of this in mind, we liked the nickname "Bugs," because at least some bugs flew. The class before ours was called "the Penguins," and as far as I knew no penguins actually flew.

That week we met with various KSC employees to learn about the significant effort that went into a launch. Administrative workers who tracked the schedule, mechanics who kept the unending list of broken equipment repaired, engineers who solved complicated mechanical problems, scientists who organized the experiments that would be launching into space, security personnel who kept us safe, military liaisons who coordinated search and rescue forces and kept the range clear of small planes and boats, support personnel who kept us fed and clothed in the astronaut crew quarters, public affairs workers who kept the press and general public apprised of our status, etc. The big learning point for me? Launching rockets, especially with people strapped to them, is a complicated business involving thousands of people.

> And then I asked, "What is that red button for?" An innocent question to be sure. Our poor guide, who was having a great time touring these neophyte astronauts, stopped dead in his tracks and the blood literally drained from his face.

We also visited the beach house, the site of many astronaut parties as well as tearful farewells, where crews said goodbye to their friends and families in the days before launch. We saw countless alligators and egrets and even a few manatees. We marveled at the cavernous VAB (Vehicle Assembly Building), one of the world's largest buildings, where the massive Saturn V had been assembled. We stood in awe at the multimillion-pound crawlers, which had moved the mighty Apollo rockets as well as modern space shuttles out to the launch pad. We saw absolutely fascinating artifacts from the 1950s and 1960s at the Air Force Space & Missile Museum at the Cape, the site of the original Mercury mission control.

We pondered the tragedies that had occurred here—the *Apollo 1* fire as well as the *Challenger* disaster. I wondered what would have possessed the managers to make the decisions they made that ultimately doomed both of those crews. I was acutely aware that NASA leaders are very smart, motivated, and patriotic people. They certainly were not consciously putting those crews' lives at risk. I wondered what kind of pressure would cause otherwise brilliant and good people to have made the decisions they made.

Myriad thoughts poured through my head: "I can't believe I'm really here," "This sure is complicated," "I can't wait to fly," etc. The emotions and ideas that fill a new astronaut's head are magnified times a thousand when he or she visits the Mecca of American spaceflight—the Kennedy Space Center in Florida.

One moment made this particular afternoon even more memorable. As we visited the various offices and departments of the Kennedy Space Center—Liquid Nitrogen Supply, Shuttle Solid Rocket Booster Parachute Rigging, Rail Logistics, Space Station Module Processing, etc.—we made our way down to the south end of the facility, on the Air Force Cape Canaveral side. We went into one of the launch control facilities that looked very cool, like something from a James Bond movie: dark, lots of computers, flashing lights, official-looking and official-sounding people, fancy seats, and titles for the various people who worked there.

And then I asked, "What is that red button for?" An innocent question to be sure. Our poor guide, who was having a great time touring these neophyte astronauts, stopped dead in his tracks and the blood literally drained from his face. He sheepishly replied, "Uh, you really don't know?"

Nope—none of us knew. But we quickly found out. You see, rockets basically launch on autopilot, though the space shuttle had a very limited capability for the crew to take over manually. So, in the extremely unlikely case of a shuttle or even unmanned rocket veering off course and heading toward Disney World in Orlando, there were some time-critical steps that would be taken. First, flight controllers in Houston would notice the trajectory deviation and call the crew along with recommended actions. In the first ninety seconds of a shuttle launch, manual flying was not possible, so the commander would engage the backup computer and hope that it would start flying in the right direction. If the off-course flight occurred after the ninety-second point, the commander could take over manually and adjust course to head out to sea, away from populated areas. If neither of those options worked—well, that's where the Red Button came into play.

First, Houston would call the crew with a secret code word, warning them what was about to happen. Next, the Range Safety Officer would take control of the situation and press the Red Button, sending computer commands to the rocket that would trigger multiple explosive devices on board, causing the rocket to split apart, avoiding populated areas and unsuspecting civilians, but also killing the astronauts on board. This decision would have to be made within seconds.

Our poor guy's face turned as white as a ghost as he apologetically told us the details of how he, a Range Safety Officer himself, had in his finger the power to send this FTS (flight termination system) signal to our vehicle that would command the shuttle to blow up, and there was nothing that we could do about it. This was good news for Disney World and the people of Florida, but not for the humans on the rocket.

This little gem of knowledge caused some awkward laughs and morbid fighter pilot jokes—"Can I have your stereo dude? Where are the keys to your car?" It was entirely understandable that the safety of civilians on the ground was more important than astronaut safety, and during my sixteen years at NASA I never heard anyone question this system. However, it did make me chuckle and wonder if this guy won the award for "most ironic duty title"—Range Safety Officer. It certainly wasn't Astronaut Safety Officer.

Red Button notwithstanding, it was a great first official trip to the Kennedy Space Center, and my belly was burning with desire to go into space after seeing *Atlantis* and her crew on the launch pad. And though the FTS system had been used over the years to terminate unmanned rockets that had gone helplessly off course, thankfully it was never used to preemptively terminate a human mission. But it was a sobering reminder for our class of the seriousness of the profession we had chosen.

Flying in space is not for the faint of heart. Or OCD control freak.

THE RIDE UPHILL

Staying Cool When You're Blasting Off

The roar was so intense. I thought I had experienced a lot as a test pilot and jet fighter pilot, having flown more than forty different types of aircraft. But that roar was incredible. I was acutely aware that a significant emotional event was about to occur in my life. I was *Endeavour*'s pilot, it was 0414 on the morning of February 8, 2010, and the three main engines had just roared to life, undergoing a quick six-second check by the onboard computers to make sure everything was working, because at T–0, when the solid rocket boosters (SRBs) lit off, there was no stopping them. We were going into space—or we would die trying.

Then night turned to day. At T–0, when the SRBs lit, 10,000 pounds per second of high-explosive fuel shot out of each nozzle, on top of the 1,000 pounds per second racing through each of the three main engines. The roar, vibration, acceleration, and sheer violence of that moment shocked me, as our four-million-pound vehicle leapt off the launch pad in an instant. The light from the rockets' fire reflected off a thin layer of clouds hovering 5,000 feet above, casting a blinding bright light for many miles. Take a minute to Google "STS-130 launch," and you'll find home movies that people posted of our launch. You'll hear people getting excited by the countdown, but they really start screaming and cheering at launch, when those clouds lit up. Their reactions are amazing, genuine, unbridled exuberance, awe, and "*We* did this!" To witness a space shuttle launch was to see firsthand the best that humanity could accomplish.

While those tens of thousands of people were enjoying our launch, nothing could have been further from my mind. I had a shuttle to fly and I had to keep my head in the game. *Endeavour* quickly rolled to aim toward the orbital plane of the space station. The main engines throttled down for about thirty seconds, then throttled back up, a ballet of rocket science and trajectory guidance designed to keep the fragile space shuttle from being squashed by the onrushing air pressure in the thickest part of the atmosphere. "Roll program, Houston" and "*Endeavour*, go at throttle up" were the radio calls between mission control and our commander, George Zamka (Zambo). Though I had heard those calls a million times in training, being smashed against my seat by the mounting g-force, overwhelmed by the roar of 23,000 pounds per second of exploding fuel, and seeing the night turn to day all made this actual launch an experience infinitely more intense than the simulator back in Houston.

One thing quickly caught my attention—that thin deck of clouds was now glowing yellow, reflecting the flames of our engines, rapidly approaching us, getting bigger and bigger, moving *fast*, to the point that I could see details in the wispy clouds as *Endeavour* shot upward at 500 mph. I winced as we punched through this bright wall in the sky, and in an instant the sky became black. Another thin deck of clouds appeared, this one up at 35,000 feet, and I had the same reaction—fascination, followed by wincing as we flew through it seconds later. Finally above all clouds, we were enveloped in blackness, rocketing away from Earth, gaining speed at a dizzying rate, climbing into an orbit that would eventually rendezvous with the International Space Station. This first minute and a half of flight was so spectacular and so far beyond anything I'd ever experienced that it was a challenge to focus on my piloting duties.

The g-forces had built up to two and a half times the force of Earth's gravity. Imagine lying on the floor, and then having a couple of your best friends lie flat on top of you, smashing you. It's not unbearable, but it's definitely not normal. Then imagine that feeling lasting for eight and a half minutes. During that time, our acceleration profile built up to two and a half g's, then tailed off to one g when the SRBs were jettisoned, then slowly built back up to three g's as we burned fuel, becoming lighter and lighter. With g-forces crushing my chest, I had to remember how to breathe. It doesn't happen automatically—you have to actively push your chest out to allow your lungs to inflate. It was a very different sensation than the nine g's I had experienced as an F-16 pilot, which had been in the head-to-toe direction. In a rocket the g-forces kept on going and going and going, whereas at least in a fighter jet they would usually stop after thirty seconds or less.

Passing the seventy-five-second point, I briefly thought about our two space shuttle accidents. It was roughly at that point that both the *Challenger* and *Columbia* accidents occurred. That brief moment is called *Max-Q*, or maximum dynamic pressure, when the vehicle is flying fast but still at low altitude. The combination of high speed and thick air causes a tremendous amount of pressure on the front of the rocket. Our airspeed was about 500 knots, and it sounded like a freight train driving past our front windows. I was shocked at how loud that roar was as the crushing force of extreme wind pushed against us. It was at this point that the *Challenger*'s SRB had vented hot gas into the fuel tank, causing an explosion. It was also at this point that a piece of foam had popped off *Columbia*'s fuel tank into the

> When the engines shut down, we were suddenly surrounded by a cloud of ice particles that had floated up from the shuttle's engines—hundreds of pounds of frozen oxygen and hydrogen, all sparkling and shimmering in the sunlight that was just peeking above the horizon over Europe.

500-knot windstream, shooting back into her wing, creating a large hole that would prove fatal two weeks later, during re-entry. I was acutely aware that this was a very dangerous moment, and there was absolutely nothing that I could do about it.

Beyond the acceleration, there was terrible vibration. It felt as if someone had grabbed me by the collar and was shaking me. I had heard that solid rocket fuel produces a lot of vibration, and they were right. Noise. Fire. Light. Acceleration. Vibration. The first two minutes of a space shuttle mission.

After the solid rocket motors had expended their two million pounds of fuel, the g-forces began to trail off, dropping down to one g, and the vibration and noise subsided. I had been watching the clock and gave the crew a polite heads-up—"Stand by for a little bang," just before the spent SRBs were to be jettisoned. That was the understatement of the century. I had a front-row seat at the window for this big event, and I looked up at the perfect time to see the three forward-facing RCS jets (NASA acronym for small rockets, or thrusters) fire for about half a second, forming a shield with their exhaust to protect our windows from the blast of the SRB jettison motors. Those motors were basically missile rocket engines that fired directly at the shuttle in order to quickly push the boosters away at separation. When they fired, there was such a loud roar and flash in the cockpit that my crewmate Steve ("Stevie-Ray") Robinson immediately answered my "little bang" warning with a "or a big one!"

Once the solids were gone, the g level dropped to a comfortable Earth-like force, and the roar and vibration stopped. It became amazingly calm and smooth, like being on the ocean on a calm day. For a few moments. As we burned fuel from the external tank, *Endeavour* became lighter and lighter, turning our liquid hydrogen and oxygen fuel into water exhaust at a rate of 3,000 pounds per second. The constant force from the three main engines pushing on a lighter and lighter vehicle meant that the acceleration built up from one to three g's. In the minute before the engines shut down, we were adding 100 feet per second every second to our velocity!

During that ride into space, I stole a quick glance outside, saw the Moon in front of us, and called to my commander, "Hey Zambo, look out the left window!" It was a few seconds of *wow* and then immediately back to work. The shuttle had launched in a head-down attitude, and upon accelerating through Mach 13 (about 10,000 mph) she rolled head-up to point our radio antenna up toward a communication satellite. The computer chose which direction we rolled, left or right, and as we did this 180-degree roll, I lucked out as we rolled to the left, giving me a view of the entire East Coast of America, from Georgia to Boston. I wasn't expecting it, but I immediately recognized I-95 and all the major cities. *Wow!* And then quickly back to work.

When the engines shut down, we were suddenly surrounded by a cloud of ice particles that had floated up from the shuttle's engines—hundreds of pounds of frozen oxygen and hydrogen, all sparkling and shimmering in the sunlight that was just peeking above the horizon over Europe. *Wow!* A few minutes later we flew into that sunrise, and the entire curve of the Earth was covered by a bright, intense, thick blue light. I thought, "I've never seen that shade of blue before." *Wow!* But I had to get back to work, preparing the orbiter for our next maneuver. The sun was finally up as we passed over the European Alps, still climbing, traveling five miles every second. The snow-covered peaks zoomed by as we crossed countless Swiss and Austrian valleys, and I thought, "When I used to live there it would take hours, and now it's only taking seconds to cross those valleys!" Yet another *wow!*

If I had to summarize the launch experience in one word, it would be: *Wow!*

There were so many things to see outside the window, and I had the most intense urge to spend time looking at them, taking it all in, photographing everything. But I was the shuttle pilot, and I had a job to do. That situation was repeated constantly throughout my seven-plus months in space. There were such sublime views and experiences, and yet most of my time was spent on mundane tasks. I quickly had to learn how to avoid distraction. The key for me was to be aware of my surroundings, and if I was tempted to stare at something for too long, to pull away from it and deliberately focus on the tasks at hand. In the fighter-pilot world, we called this "task saturation," and it was often deadly. Staring at the target or instrument or sensor for too long almost killed me on several occasions in jets. There was the same temptation in space, except seeing our planet from space is *way* more enticing than staring at a dial in an airplane cockpit. But the results could be equally deadly. As a pilot or astronaut, as in many other professions, you can't let yourself get fixated on one task. That's a recipe for disaster.

The ride into orbit is the most exciting, utterly intoxicating eight and a half minutes you can imagine. But the harsh reality is that you have to be disciplined when experiencing launch. Because target fixation is a bad thing, whether you're flying an F-16 or a space shuttle.

ORBIT

LEARNING TO FLOAT

How to Cope with Zero G

think Pink Floyd needs a follow-up song. "Learning to Fly" was great, but "Learning to Float" would really be a hit. A lot of people have learned to fly since Orville and Wilbur Wright first took to the skies. But learning to float, that's an entirely different ball game. And that's precisely what you must do when you ride a rocket into space and the engines shut down. Because from that moment until your spaceship returns to Earth, often half a year later, you are floating. And you don't have a long time to learn—you need to figure it out ASAP. With no grace period. The main engines cut off and you are floating, weightless, and you better have a plan, or you will be flailing like a fish out of water.

"What does it feel like in space?" has a simple answer: It feels like you are falling. Because you are falling. You see, at the altitude of the space station's orbit, Earth's gravity is still very strong, roughly 90 percent of what it is on the surface of the planet. The difference is that you are moving. *Fast*— 17,500 mph (8 km/sec). Every second, you drop toward the center of the Earth about 15 feet (5 meters). However, during that second you move forward 5 miles (8 km). Then the next second you drop another 15 feet, but you also move forward another 5 miles. And so on and so on. That motion makes a curve, and if you are at precisely the right speed, that curve will match the shape of the Earth. Et voilà—you're in orbit, staying roughly the same altitude above the Earth's surface.

The sensation of falling is one that every human has genetically wired into our brain—we intuitively know that if we are falling, something very bad is about to happen. That is probably why we flail our arms, maybe trying to fly, maybe trying to grab on to something on the way down to the ground. But in space, flailing does you no good at all, trust me. So rookies need to deliberately think through the experience of weightlessness *before* they get to space. Consciously don't flail. Consciously don't kick your feet. Consciously move slowly and deliberately. These very basic skills are important because you can hurt yourself and your crewmates if you don't abide by them, and many rookies have done just that. If you look at video footage of first-time flyers, or space tourists, you'll see scratches and cuts, especially on their first few days in space. There are even stories of folks getting punched or kicked by accident as a new guy moved around carelessly in the unfamiliar environment of space.

One of the most shocking things about being in space was the speed and randomness with which stuff floats away.

The first skill to master is to move slowly. If you move too quickly, you're going to bang your head into something or end up floating to the wrong destination. What's worse, you'll spin yourself around on the way. So you push off slowly and deliberately, aiming directly toward your destination. On Earth your brain knows to compensate for gravity, so when you throw a ball to someone you aim high. The same mental compensation happens in space, except the ball doesn't drop, so for my first few weeks in space I was constantly throwing tools to my crewmates above their head. It takes time and deliberate thinking to float in a straight line directly to the target, not high.

The next problem is rotation. If getting yourself or some object to the correct destination in space isn't hard enough, try to do it without spinning it around. There were several very humorous instances on my first flight when I pushed off to fly across the module, and I aimed myself very well, hitting the destination. Except I began to spin immediately after pushing off, floating helplessly between two walls, my body slowly rotating like a chicken on a rotisserie spit, hitting the destination back-first. I always tried to look cool 'cause I'm a fighter pilot, with a "Yeah, I meant to do that" look. Only I didn't mean to do that. The physics that my brain had learned on Earth, in terms of how things fall or bounce off each other, had now transitioned to pure Newtonian physics, $F = m \times a$ (force = mass × acceleration), with no friction or gravity. It's a different, even alien environment, but boy is it fun once you learn how to float.

Handling your body is only half the problem; you also have to learn how to handle things, and that is an entirely unique skill of its own. One of the most shocking things about being in space was the speed and randomness with which stuff floats away. If you have a checklist, or pencil, or batteries, or whatever, and you leave them unsecured, you will lose them faster than a tourist in Rome can lose a wallet from his back pocket. Intellectually, I knew that there wasn't someone sneaking around making those things disappear, but at times it felt like an invisible force was causing them to fly off.

The most entertaining phase of this learning curve was the first few days in space. Nothing is better than watching a rookie (myself included) getting dressed in the morning. I would bounce off the walls while holding my shorts—which were stuck around my feet, with my shirt floating away, etc. Eventually, I figured out how to properly restrain myself and manage clothes, usually by wedging my feet under a handrail and my clothes under a bungee, pulling them out piece by piece as needed. That deliberate process of keeping things restrained until they were needed worked well for just about any task.

There are several methods to restrain items in space: pockets, Velcro, tethers, Ziploc bags, bungee cords, straps, storage bags, and containers. Those are all seemingly innocuous or even extraneous things down here on Earth, but they enable work to happen in space. Without them, the ISS would be filled with a giant cloud of mess and you could never get anything done. Imagine what would happen if your teenager's room suddenly lost gravity and everything in it began to float. That's what a spaceship would be like without some handy Velcro and Ziploc bags.

There is one very cool exception to this rule—you can float something in front of you, steadying it before releasing it, and it will stay put for a short time. This came in handy when doing maintenance. I would float a wrench or screwdriver while I quickly used both hands to finish another task, then grab it as it floated next to my head. Once you got really good at this, you could float something in the middle of a module as your crewmate was on his way, and he could grab it out of midair on his way to his destination. Kind of like Amazon Prime delivery in space—very fast and convenient. You just had to keep an eye on stuff you were floating and not forget and leave the area, because then it would be gone.

One of my more famous lost-item episodes involved a small camera. NASA sent up three Ghost-brand cameras, small video/still cameras similar to a GoPro. Being a camera guy, I opened them up as soon as they arrived on the SpaceX Dragon cargo ship and commandeered one for myself, keeping it in my crew quarters so I could easily grab it and shoot without having to hunt for it. One of my crewmates also took one, and we put the third in storage. Literally the next day I heard from them—"Terry, I lost the Ghost." Ugh. I figured it would turn up at some unexpected time, as things normally do, almost always making their way to one of the many filters that were constantly sucking air.

Well, days turned to weeks turned to months. No Ghost. I was bummed, and felt responsible, because I was the camera guy and I had decided to just keep the cameras in our personal sleep stations instead of in their designated storage locker. It was getting near the end of our mission, so I decided to call Houston and fess up: I told them that a Ghost was MIA. They weren't too mad; there wasn't much to do since it wasn't a critical piece of equipment and we had two others. But losing something really, really bugged me and my OCD. Literally the very next day I was in Node 2, where our crew quarters were located, arranging some equipment underneath a little-used shelf, and voilà—a Ghost camera was innocently floating there, stuck to an obscure piece of Velcro on the wall, where it had hidden for more than three months.

In another instance our crew lost a torque wrench, a fairly big tool— probably a foot in length and several pounds of mass. Gone. We all looked for that thing for months. Filters. Nooks and crannies. Misplaced tool containers. Nada—it was gone. Then magically someone found it floating in a module one day. We had no idea where it had been hiding or what adventures it had gone on, but like the prodigal son, it was lost and then it was found.

In yet another instance, a few weeks after arriving at the ISS I was taking some pictures in the Cupola, a seven-windowed observation module that has amazing views of Earth and the universe. I had taken some particularly awesome shots of the aurora borealis and I was *really* excited to download them and see them. I took the memory card out of the camera to put it in my pocket, and it slipped out of my hand. Off it floated, spinning end-over-end like a quarter flipped at the beginning of a football game. And I watched in slow-motion horror, yelling "Noooooooooooo" as it slipped directly into a crack on the Node 3 wall. It was only a fraction of an inch wide, and that card floated right in there. I waited for it to bounce out and spent weeks checking filters and warned all my crewmates to keep an eye out. Nothing. Someday, decades from now, when the ISS finally drops into the Pacific Ocean, there will be a memory card holding some sweet pics of auroras stuck behind one of the walls. Sigh.

As mentioned before, astronauts experience something we euphemistically call "space brain." We performed several experiments to quantify this phenomenon, and one in particular was called Cognition. I did this test roughly once per month while in space. I would strap myself down in front of a laptop and do memory, pattern recognition, and reaction tests. The results showed that my brain did slow down a little while in space. This may have been due to floating and its resultant disorientation, or from fluid shift, or neuro-vestibular disorientation, or even elevated levels of carbon dioxide in the atmosphere that impeded brain function. Whatever the scientific explanation, I'm pretty sure that every astronaut would agree—space brain is a thing. You forget easily, don't notice things that are happening, and can't think of solutions as well as you could back on Earth, especially in the first few days after arriving in space. I've heard women describe the same symptoms while they were pregnant. And though I'd never dream of saying, "I understand what it's like to be pregnant," I do think that astronauts get a taste of mild cognitive impairment, aka space brain.

Learning to float and work and live in space was one of the highlights of my career. I loved going through that steep learning curve, until after a few weeks I was a pro—a real spaceman. Able to push off and get to my destination quickly and without spinning around. Able to keep track of tools and equipment and clothes effortlessly. Usually. Except for that memory card.

HOW TO BUILD A SPACE STATION

A Painstaking, Piece-by-Piece Process

When you see Hollywood space movies like *2001: A Space Odyssey*, *Interstellar*, or *The Martian*, there is usually some giant spaceship that magically appears and flies off to other planets or galaxies. Which makes for an interesting movie. But in real life, those beasts have to be built; they don't just appear out of thin air, or out of thin vacuum for that matter (pun intended). This chapter is the story of how the ISS (International Space Station) was built, piece by piece, over several decades, by a multinational partnership.

The idea for an American space station began during the Apollo era in the 1960s. As the Moon landings were drawing to a close, NASA was left with a lot of hardware and talent and wanted to find something productive to do with it. The AAP (Apollo Applications Program) was implemented to develop those ideas, and two missions were eventually flown. I was not even a year old when the AAP office was formed, but it would eventually affect my life in the most profound ways.

The first AAP mission was Skylab, the first American space station in orbit around Earth. Skylab was an outfitted upper stage of the Saturn rocket, and as such required only one flight to launch. Three crews visited Skylab between 1973 and 1974, with the longest stay approaching three months. The second AAP mission was the Apollo-Soyuz Test Program in 1975, in which an American Apollo capsule and a Russian Soyuz capsule docked together in Earth orbit, with the astronauts and cosmonauts performing a very symbolic space handshake in the middle of the Cold War and the aftermath of Vietnam.

While NASA was busy with AAP and developing the space shuttle, the Soviets flew the world's first space station, Salyut-1, in 1971. Several follow-on Salyut stations were flown until 1986, when the Mir space station was launched. Mir was the first true modular space station to be assembled in space, component by component, eventually comprising six modules. It was abandoned and deorbited into the Pacific in 2001, which coincided with the reason NASA hired me to be an astronaut in 2000—it was time to build the International Space Station.

As the space shuttle program got underway, NASA needed to expand its mission. Launching satellites and performing classified Defense Department missions was great, but not enough for such a high-dollar program. In 1984 President Reagan proposed Space Station Freedom in his State of the Union address, and over the next fifteen years the details of that station would change considerably.

The most important change came with the demise of the Soviet Union. President George Bush (41) directed his vice president, Dan Quayle, to engage the Russians as partners on the fledgling space station. The fear was that the collapse of the Soviet Union would lead to massive turmoil in their aerospace sector, and cooperation on a large-scale space project would provide a productive outlet for the thousands of newly unemployed Russian scientists, engineers, and technicians. It would certainly be better to have them building a space station with us than building nuclear weapons for some unseemly characters on the world stage. A partnership of sixteen nations was formed: Canada, Europe (eleven nations total), Japan, Russia, the United States, and Brazil, which eventually left the ISS partnership for budgetary reasons.

Construction of the ISS began in November 1998, when the Russians launched the FGB, or *Zarya* storage module, on a massive Proton rocket from the Baikonur Cosmodrome in Kazakhstan. Two weeks later the space shuttle *Endeavour* added the Node 1 module *Unity* to the FGB, and the ISS assembly sequence began. Over the next twelve years, the ISS was built, piece by piece, module by module, by a handful of Russian and nearly thirty US space shuttle launches.

There were two challenges to the successful completion of the ISS. The first was technical. Those modules had to be built to common specifications and fit together, built in factories around the world, and then often assembled for the first time in orbit. I used to marvel when, after a module sat collecting dust for years in the processing facility in Florida awaiting its launch, a shuttle would carry it up and attach it to the ISS, and it would work perfectly. This happened time and time again for more than a decade. The technical prowess that went into the space station was extremely impressive.

Second, and even greater than technical, were political and fiscal challenges. Keeping sixteen nations focused on the task of building and operating the ISS for several decades has been the highlight of US foreign policy in the post–WWII period, since the Marshall Plan. Each one of those sixteen nations has an electorate, domestic priorities, competing strategic partners and competitors, fiscal realities, and changing political landscapes. In short, it's a miracle that the station partnership was able to hold together over such a long period of time. The space station program famously survived the US House of Representatives in 1993 by one vote, 216–215, as well as a US presidency that has swung from Republican to Democrat to Republican to Democrat back to Republican over its thirty-year-plus history. Russia has undergone a traumatic transformation, from the collapse of the Soviet Union in the turbulent 1990s under Boris Yeltsin to Vladimir Putin. Europe, Canada, and Japan have all experienced dramatic shifts in domestic politics. The ISS has survived, even thrived, through it all.

The technical aspects of building a spaceship in orbit are fairly straightforward. Space shuttle payloads were limited to a certain size and weight, so all of the ISS modules had to be built within that constraint—60 feet long, 15 feet in diameter, and no more than 40,000 pounds mass. Had we had a heavy-lift rocket, the ISS could have been built in much less time and at less expense. There was a proposal in the 1990s for a Shuttle-C, or cargo version of the shuttle, that would have replaced the orbiter with a giant cargo module strapped to the side of the orange fuel tank and white boosters. Assuming an unmanned rocket like this had been developed with an 80-ton lift capacity, it would have taken only five or six launches to build a station with roughly the same capability as today's ISS. And at the end of that hypothetical assembly, we would have had a heavy-lift rocket for flying to the Moon or Mars or beyond. Heck, we could have just restarted the Saturn V program. Alas, we never pursued that strategy and ended up building the entire ISS with a handful of Russian Proton rockets and an awful lot of space shuttle missions—one 15-ton module at a time, each one having its own redundant and inefficient hatches, berthing systems, and extra electrical equipment to allow it to function on its own. Building a lesser number of more massive modules would have really cut down on duplication. But the space shuttle worked, and selfishly, I absolutely loved flying it.

Each assembly flight had a similar profile. The shuttle crew would be highly trained on all of the specifics of that module—how to operate it from inside as well as perform spacewalks outside. The shuttle would launch to an orbit behind and below the partially built ISS, gradually catching up over the course of two days. The commander would dock the shuttle and the crew would immediately go to work, like a NASCAR pit crew when their driver pulls in for a tire change and fuel-up. The module would be pulled out of the shuttle using its robotic arm, handed over to the larger station arm, and then attached, or berthed, to the ISS using a series of mechanical hooks and bolts. Spacewalks would be performed to plug in power and cooling cables—and take a few cool astronaut selfies. Then the shuttle crew would shake hands with and say goodbye to the station crew. They had been happy to see the shuttle guys when they showed up a week before, but they were inevitably even happier to say goodbye, like an in-law visit. Next, the shuttle crew would close the hatch, undock, perform a fly-around (a 360-degree, 400-foot-radius loop around the station), fly away, and return to Earth two days later, after inspecting the shuttle's heat shield prior to entry.

That sequence was repeated several times a year for more than a decade, assembling the station like a Lego set, until finally, on STS-130, we installed the final two modules in the official assembly sequence—Node 3 (aka Tranquility) and the Cupola. These modules grew the station to a final size of nearly a million pounds of mass and a width of more than 300 feet, bigger than a football field. There have been a few other add-ons to the ISS since that time, notably BEAM, an inflatable module test-bed built by Robert Bigelow, and the PMM, a storage module left behind after STS-133 as a permanent closet for the station. The Russians also have a few new modules on the books, though it's tough to say if and when they'll actually make it to ISS.

International Space Station (ISS)

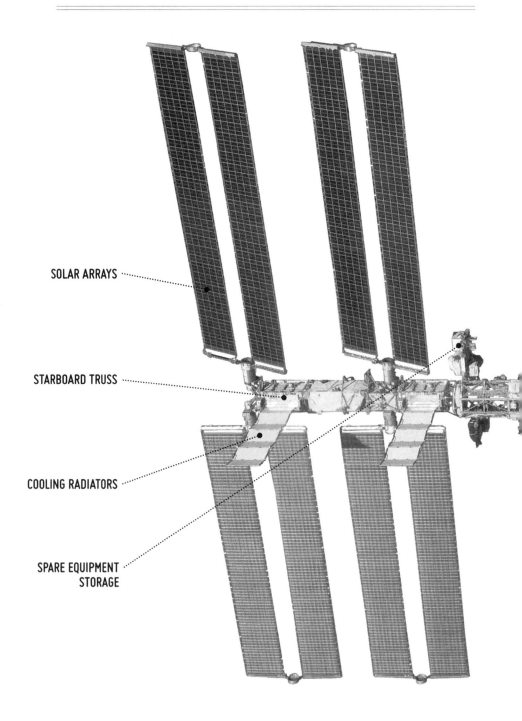

SOLAR ARRAYS

STARBOARD TRUSS

COOLING RADIATORS

SPARE EQUIPMENT
STORAGE

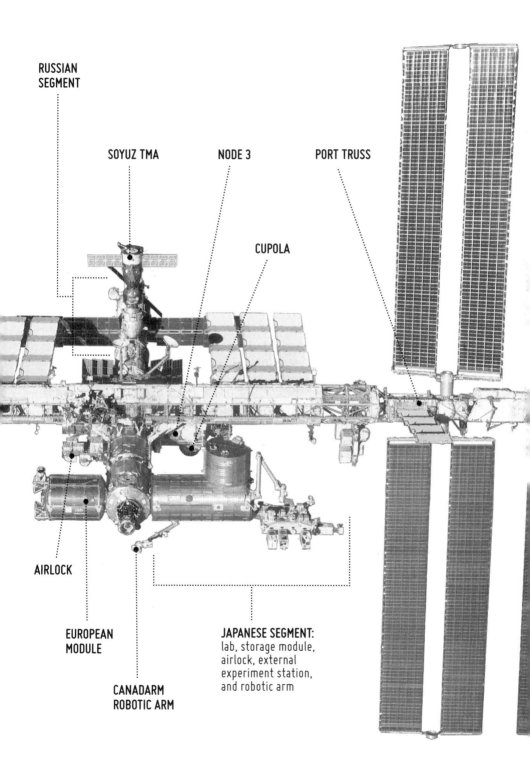

RUSSIAN
SEGMENT

SOYUZ TMA

NODE 3

PORT TRUSS

CUPOLA

AIRLOCK

EUROPEAN
MODULE

CANADARM
ROBOTIC ARM

JAPANESE SEGMENT:
lab, storage module,
airlock, external
experiment station,
and robotic arm

This piece-by-piece, step-by-step technique to build a spaceship may be used for future vehicles. NASA is planning to build a mini-ISS in orbit around the Moon, called Gateway. We will probably eventually use a similar technique to build larger spaceships that will take us to Mars and beyond. Although it is always more efficient to use smaller numbers of larger modules launched by a few heavy-lift rockets than a lot of smaller pieces launched by smaller rockets, I was a fan of the ISS assembly strategy. Selfishly. Because I had the privilege of taking part in one of the most incredible space adventures ever—flying a space shuttle on a mission to finish building the ISS. We've never built anything else so magnificent in space, and I doubt we ever will.

PILOTING SPACESHIPS

Rendezvous, Docking, and Avoiding Space Junk

lying an airplane is an entirely different proposition than flying a spaceship. I had been a pilot for twenty-five years before I first flew the shuttle in space, so I had a lot of "baggage" to unlearn. The things that make you go faster, slower, up, or down in orbit are entirely different from flying a plane in the atmosphere. So you have to forget how you learned to fly jets if you want to become a good spaceship pilot.

Before we dive into the specifics of flying rockets, here's a quick primer on flying vehicles on Earth. Every airplane has a throttle and a control stick, or yoke. The throttle is simple enough that even a pilot can't mess it up. If you move it forward, you go faster. Pull it back and you slow down. The stick or yoke, if you're unlucky enough to not be in a fighter jet, makes the airplane pitch up (trees get smaller) or pitch down (trees get bigger) and roll left or right. If you want to yaw the plane left or right, you step on the rudder pedals. Pretty simple. You are now ready to begin taking private pilot lessons. However, flying in space can be entirely nonintuitive.

It all starts with orbital mechanics. As a fighter pilot, I used to say, "Isaac Newton is in charge" when describing something that was free-falling without power, a nod to his famous $F = m \times a$ equation. That pretty much sums up how objects behave while in space. The first big difference is that you are traveling at a tremendous speed in orbit, which has serious implications for your ability to turn to the left or right even a little bit. Turning requires a tremendous change of velocity, or delta-v, which requires a corresponding amount of rocket fuel, which requires a correspondingly large fuel tank, which ultimately requires a lot of money. A rocket scientist would call moving left or right a plane change, or changing your inclination, which is the heading as you cross the equator. A spaceship continuously flying over the equator has an inclination of zero degrees, but the ISS orbits at an inclination of 51.6 degrees, so it crosses the equator 51.6 degrees north or south of pure east. Your inclination also happens to equal your maximum latitude, so in the case of the ISS it is at 51.6 degrees north or south latitude 22 minutes after passing over the equator. It is back over the equator another 22 minutes after that. Because rockets have a very limited ability to change inclination, they are basically stuck with their launch inclination. So, if you find yourself flying a spaceship in the near future, be sure to launch on the correct heading, because you won't be able to make significant changes to the left or right.

The next fundamental concept is that satellites move more slowly in higher orbits than in lower orbits. Unlike an airplane, where pushing the throttle forward makes you go faster, accelerating a spaceship first speeds you up, which makes you climb, which then slows you down. Not intuitive, but it's what happens. In the same way, if you initially slow down, you will sink to a lower orbit, which ultimately makes you move faster. Let's put this into practice. If you want to close on an orbiting object in front of you, first slow down; then you descend without doing anything else, which then speeds you up and you begin to close on your target. This principle can also be seen in the orbits of the planets. It takes Mercury only 88 days to go around the sun, whereas it takes Pluto (I'm old school and Pluto is a planet!) 248 years, because Pluto is in a much higher orbit, farther from the sun.

A practical application of this principle can be seen with geosynchronous satellites. They are very high up—more than 22,000 miles above the Earth, where it takes them 24 hours to complete one orbit. What's more, if one of those satellites has an inclination of zero degrees it will orbit directly above the equator. So a satellite 22,000 miles above the equator will stay in that same position over Earth, which is handy for satellite TV coverage—you can point your dish in the same direction and the satellite will always be there.

We had a very unexpected but graphic demonstration of this principle while on Expedition 43. My crewmate Samantha Cristoforetti was doing a personal experiment for a physicist friend of hers, who was investigating the behavior of particles in zero g. He wanted to see how clouds of particles interacted with each other in weightlessness, which could help illustrate how planets and solar systems were formed. To do this, she built a highly sophisticated device—a clear plastic ball filled with breath mints and M&M's—to simulate the primordial solar system. She would shake it up and film the small objects as they bounced off each other in random directions. After a few days of this, we noticed that all the candy ended up in a pile on one side of the ball, which made no sense; we had intuitively expected them to continue bouncing around in random motion.

Unbeknownst to us, we were seeing a demonstration of how everything on the ISS was in a slightly different orbit, if only a few millimeters apart. And objects orbiting at different altitudes should move at different speeds. However, because the candy at the top of the ball was in a higher orbit than the ISS center of gravity, if only by a few feet, it should have been moving a little bit slower than the rest of the station. Because it was moving faster than it wanted to, it tried to climb, which caused it to move to the top of the ball, above the center of the station. This was wholly unexpected and fascinated the entire crew. I took that ball and placed it in different modules, and in each case the M&M's floated away from the station center, stabilizing after just a few minutes. This demonstrated an effect known as gravity gradient, which causes spacecraft during rendezvous to naturally fall away from their target, and also causes elongated objects to naturally float up and down, with their long axis pointed down to Earth. It was an unexpected but fascinating physics lesson, thanks to Sir Isaac Newton!

As a space shuttle pilot, it took me some time to master these basic principles, but after a while they became second nature. During *Endeavour*'s rendezvous with the ISS, I was charged with performing a few small rocket burns during the final hours before docking. Those fine course corrections helped us approach the station on a precise, predetermined trajectory from behind and below, eventually ending up directly below the ISS. At that point, our commander, George Zamka, took over and manually flew the rest of the approach, eventually looping around to the front of the station, then slowly backing in, at the speed of paint drying, until we gently docked with the front of the ISS.

Space Shuttle (at launch)

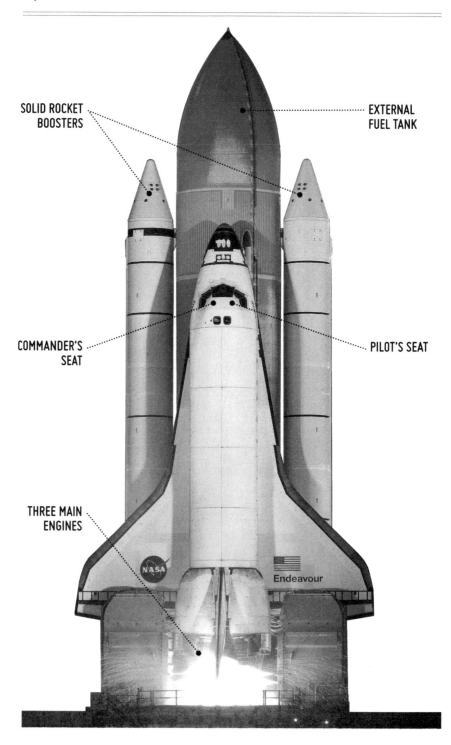

SOLID ROCKET
BOOSTERS

EXTERNAL
FUEL TANK

COMMANDER'S
SEAT

PILOT'S SEAT

THREE MAIN
ENGINES

NASA

Endeavour

Space Shuttle (in orbit)

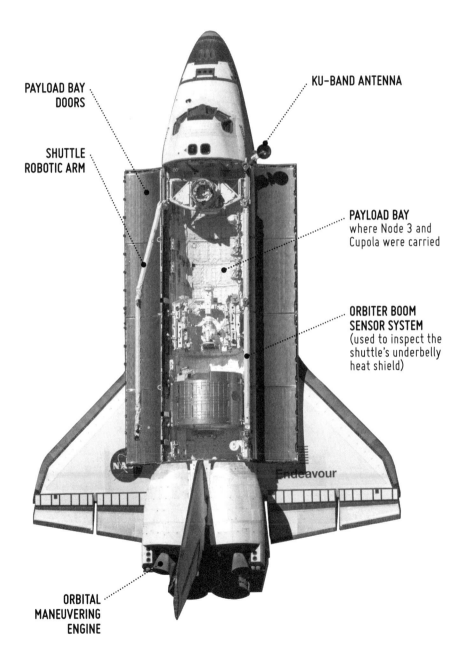

PAYLOAD BAY
DOORS

SHUTTLE
ROBOTIC ARM

KU-BAND ANTENNA

PAYLOAD BAY
where Node 3 and
Cupola were carried

ORBITER BOOM
SENSOR SYSTEM
(used to inspect the
shuttle's underbelly
heat shield)

ORBITAL
MANEUVERING
ENGINE

Ten days later, I finally got my chance to fly *Endeavour* for undocking. As she slowly moved away from our docking port, I had to make continuous down inputs to the control stick to keep her from climbing as we gradually accelerated in front of the station. Then, while flying a giant 400-foot-diameter loop around the station, I made a constant stream of small rocket firings to keep us on a circular path, constantly fighting Mr. Newton, who was trying to pull us away, or accelerate us or decelerate us, depending on whether we were above or behind or below the ISS. I'll never forget one moment when we were directly above the ISS, flying over the Himalayas on a bright clear day, looking down at snow-capped Mount Everest. There was a sea of look-alike peaks below, and it was breathtaking. The entire crew was jammed onto the crowded shuttle aft flight deck, pressed against the windows, enjoying the spectacular view. I basically flew that whole maneuver with my arms and body wrapped around the controls to make sure nobody bumped them while clamoring for a photo. Luckily the whole maneuver went well, and we all made a lifetime of memories during that brief loop I flew around the station.

The same orbital mechanics that were required to properly fly a rendezvous or undocking in space were also required to avoid other objects in orbit. The US Air Force (and now the US Space Force) tracks tens of thousands of man-made objects that are orbiting Earth. Some of them are big, operational satellites. Some of them are dead satellites that have lost their ability to maneuver after years in the harsh environment of space. Some are spent rocket boosters, drifting in highly elliptical orbits after lofting their payloads into orbit. And the most dangerous are small pieces of space junk, leftover parts of launch vehicles or clouds of debris from accidental satellite collisions.

The worst kind of debris is that from military antisatellite weapons, where a satellite is intentionally destroyed by an interceptor. Those actions can create thousands of small pieces, all of which are potentially lethal to other satellites in orbit, or worse, humans on the ISS. In fact, during my tenure on the station, we had to maneuver to avoid an object from an anti-satellite demonstration that the Chinese had performed way back in 2007. That explosion created a cloud of debris that will be in orbit for many decades and causes the ISS to maneuver several times a year, at a cost of millions of dollars in propellant and lost crew time. Following that test, the United States launched an antisatellite weapon in a tit-for-tat maneuver to demonstrate that we had the same capability, but the target of the American test was in a much lower orbit and its debris burned up harmlessly in the atmosphere a few days later, posing no threat to other satellites or humans.

Unfortunately, India joined the ASA (antisatellite) club in 2019 by destroying a target in low Earth orbit, which created yet another cloud of debris, like a shotgun shooting a clay pigeon. In and of itself this was an irresponsible event. In a larger context it reminded the world of the Kessler syndrome, a theoretical cascading effect in which one satellite explosion results in a cloud of debris that destroys other satellites, snowballing into a much larger cloud of debris that essentially ruins a significant block of orbital altitudes for any satellite. Such a disaster—many thousands of small parts of exploded satellites zooming around low Earth orbit at 5 miles per second in random directions—would remain for years if not centuries and would create a mess that we would not be able to clean. A reckless (or deliberate) actor could quite literally render low Earth orbit unusable. Imagine a world without GPS. Such a scenario is quite possible if we do not treat space responsibly, and it is a major reason for a renewed focus on the importance of space security. The community of spacefaring nations as well as spacefaring private companies needs to get its act together to ensure this space-debris Armageddon doesn't happen. We need to take care of our environment on Earth, and also above Earth.

The Air Force tracking network gives NASA periodic updates on predicted collision risk. Several times per week the flight controllers in Houston are notified of potential threats, and those warnings almost always result in the conjunction (point of closest approach) resolving to a green status, with no action required. However, they occasionally end up as yellow or red. You can guess the meaning of each of those. If the debris is in a stable and well-known orbit, and we are given enough of a heads-up, we can plan what NASA calls a DAM (debris avoidance maneuver). The station's orbit is adjusted, usually by climbing, to create enough miss-distance to make the conjunction status turn green. These maneuvers are performed by the Russian Service Module rocket engines in the back of the ISS. Imagine a Hollywood movie depicting such a maneuver; astronauts would be yelling in their spacesuits while being smashed against their seats by unimaginable shaking and acceleration and terror—with maybe a few aliens thrown in. Reality is slightly different. I have a great video of me during our DAM; I was in workout gear, slowly floating by, inch by inch. It's a funny clip.

A question that I'm often asked is, "Can you see other objects flying around in space?" The short answer is no—not usually. Those pieces of debris and other satellites are moving at many miles per second and are usually hundreds or thousands of miles away. However, at night you can actually see them in the distance, if they are high up in sunlight and the ISS is in darkness. You can also see them from Earth during the same conditions. Find a dark place, far from city lights, look up in the sky about an hour after sunset, and you'll see small white points of light moving slowly, not blinking like aircraft do. I usually didn't notice them until I was reviewing a time-lapse movie that I had shot, though I did occasionally see them in real time with my own eyes. I even saw a few meteors, always below the station, burning up in the atmosphere. You can also see those from the ground, especially during the annual Perseid or Leonid meteor showers. But when you're on Earth you need to look up, and when on the ISS you need to look down!

When I was picked to be a shuttle pilot, I knew that orbital mechanics were different than flying airplanes, but it wasn't until I went through rendezvous training that I realized the extent of that difference. A few years later the concepts of orbital maneuvering had really clicked in my brain, and I could fly my spaceship intuitively, a transformation of me as a pilot that I had never anticipated, but which was very welcome. Sir Isaac would be proud.

JUST ADD WATER

Space Station Cuisine

A s they say, "Keep the crew well fed and they can survive anything." And I think they're right. Any mission longer than a few hours should involve food, and if the food is good, the crew will be happy. If it's not good, it's going to be a long mission. This was true for both of my spaceflights, and I was a happy crewmember for my seven months in space.

The days of astronauts eating food paste from a tube are long gone. Even though we don't exactly have Michelin-star cuisine in space, I found it to be good and a highlight of my missions. For that matter, how often do you eat in a Michelin-star restaurant on Earth? The most important thing for me, and for most of my colleagues, was variety. The second most important thing was having enough to eat. Doing two and a half hours of exercise every day in addition to being extremely busy meant that I was hungry and needed to eat. Thankfully, there was plenty of food for the whole crew, though there have been several missions during the ISS program when food started to get scarce because of resupply vehicle problems, and that led to some crew hunger.

There are several categories of space food. The first is thermo-stabilized, which means it can last for months and is ready to eat without rehydration, like the MRE (military rations). This food came in green bags and included meat, vegetables, desserts, soups, etc. You simply opened the bag (reheated if desired), and it was ready to go. Another kind is rehydratable food. This type was lighter, smaller, hard, and crunchy and came in clear, see-through bags. You would plug it into a machine, select how many milliliters of water you wanted to add, press the blue (ambient temp) or red (hot water) button, unplug the food, spin it around like a centrifuge, mash the wet bag of food around for a minute, let it sit for ten minutes, and voilà, food—meats, vegetables, fruits, desserts, basically anything that can be dehydrated. This food also tasted good, and it seemed fresher than the green-bag food. These two types were the bulk of my diet while on the ISS.

Our chocolate situation will likely go down in the annals of exploration history as a tale of stretching both rations and human endurance to their limits.

There was also off-the-shelf food: M&M's (called "candy-coated chocolates" in NASA's awkward alternative universe). Tuna fish in a bag. Spicy olives. Chocolate-covered blueberries. Reese's peanut butter cups. Power bars. You get the idea. Any food that can be purchased at the local grocery store, stay fresh for a few months, and be shipped "as is" to space can probably find its way onto a cargo manifest. In terms of bread, we have only tortillas on the American segment, because they can last a long time without spoiling or making crumbs like a normal loaf of bread would. Fresh fruits and vegetables, like oranges, carrots, and apples, are a welcome treat on newly arrived cargo ships. Oranges tended to rot pretty quickly, but they sure did smell good. It was always amazing to have fresh food after eating packaged food for months.

There were also drinks. Any type of drink that can be produced in powder form can become a space drink, including fruit juice (yes, it's actually Tang), tea, coffee, milk, sports drinks, hot chocolate (in short supply and high demand), and even smoothies. There was a pretty good selection of everything other than carbonated drinks. I was a Diet Coke addict before my flight and was nervous about going without for half a year, so I requested tea, figuring that would be my caffeine drink. Unfortunately, after about three days of drinking tea I had had my fill. There were lots of varieties of tea: black tea, green tea, sweet tea, tea with lemon, etc. But the bitter taste was too much. It was great once every few days, but not every day.

The vast majority of our food was organized in BOBs (NASA acronym for food bag) that each contained eight days of that particular category of food for the whole crew. They are fairly dense containers, the size of a backpack, and are organized by type of food—meat, vegetables, desserts, drinks, fruits/

nuts, breakfast, etc. The contents are part of a standard menu of food, which is selected to appeal to most astronauts. We were supposed to go eight days between new BOBs for most categories, though we would usually last longer than that.

Opening a new food container usually took about fifteen to twenty minutes by the time you went to find the new one and scanned it into the inventory management system, so it was always a bummer if you were the last person to finish off a BOB and had to go restock the cabinet. It was a job that we all shared, without formally being tasked; if we saw the need to open a new BOB, we would just do it. That attitude was very helpful for crew cohesion. If I saw a crewmate restocking food supplies and I had a minute, I would always stop and help out.

Beyond standard menu items that the whole crew shared, each of us was given nine personal bonus containers of food. These could be from the NASA, Russian, European, or Japanese menus or even items from the local grocery store. I picked mostly US, but also a few containers of Russian and European food. The key was variety. The Russians were particularly good at fish, mashed potatoes, and soup. Plus they had real bread—*Borodinskiy khleb*

Drinking a space milkshake: Ben and Jerry's ice cream, powdered milk, and recycled water mixed in a Ziploc bag. Messy, but yummy!

(бородинский хпеб, or Borodinsky bread), a dark rye sourdough that was basically the only bread I ate for 200 days. The only European food I was able to get was some leftovers from previous European astronauts. They had dishes made by real chefs, and those were always a treat.

We would also get care packages from Earth with each visiting cargo ship, usually containing beef jerky and chocolate candy. It was kind of funny because we were overrun with jerky, but for me the chocolate was most important. So when the Cygnus cargo ship blew up just before my launch, and with it one of my care packages full of Reese's peanut butter cups, I was a little alarmed. Then, halfway through my mission, a Russian Progress cargo ship blew up, and along with it more chocolate. Our mission duration was then extended indefinitely as a result of that accident, and I had to start rationing my chocolate. It was painful. Our 169-day mission became a 200-day mission, and I literally ate my last Reese's on flight day 200, hours before leaving space. I had barely made it. Our chocolate situation will likely go down in the annals of exploration history as a tale of stretching both rations and human endurance to their limits.

The food setup was a little different on the shuttle; there we chose every meal for the whole flight. It was a good system in that we ate what we wanted, but it made it less likely that you would try other food items. Like many astronauts my taste changed while in space, but the shuttle flight was so short and I was so busy that any lack of variety didn't matter. At the end of STS-130, we left a huge bag of supplies on the station, including food, which was a morale booster for the ISS crew remaining behind.

One of my favorite food stories involves an experiment I did called Astro Palate. It was a psychology experiment designed to measure how food affects our mood. They had me eat bland food and then do certain mundane or unpopular tasks, like Saturday cleaning or repairing the toilet. Next week I would eat some of my favorite dishes and do the same tasks. Some weeks I would eat the food after doing the tasks. Then I took a survey of my mood both before and after the tasks. Sure enough, I was a lot happier after eating the good stuff. Samantha gave me endless grief about this experiment because while she was busy giving blood or doing other invasive medical tests, I was eating chocolate brownies after vacuuming filters. We laughed a lot about that, but I never felt badly about it. Unless I had to do the cleaning after eating cheese grits.

As you might imagine, there were some food items that weren't favorites. I was lucky because my crew had pretty diverse tastes, and it would be a bummer if everyone wanted to eat the same food. However, a few things were always in demand—brisket, shrimp cocktail, chocolate brownies, scrambled eggs, sausage, hot chocolate, and, ironically, veggies. Nobody on my crew liked cheese grits or curried vegetables or most of the thousands of packets of tea. So I started a bag of uneaten food, a large bag of food items that the US segment crew (Americans and Samantha) didn't want. Every few weeks the Russian cosmonauts would come down and raid that bag; they really enjoyed the food that we didn't want. Conversely, we loved the extra items that they were tired of, especially canned fish. This system of sharing food worked great; everyone had good variety, nothing went to waste, and I don't remember throwing away any food during my mission.

I was very thankful to the NASA, Russian, and European food labs for making some very good food for us. Between exercising two and a half hours a day and no fried fast food, I came back to Earth trim and in great shape. I still remember the first meal I ate back on the planet—one of my doctors went to the airport kiosk and got a chicken sandwich for me. The fresh bread and mayonnaise were absolutely amazing.

Even though our space food doesn't compare to a freshly prepared meal here on Earth, some of my best memories in space occurred around the dining room table, food velcroed and duct-taped to the table, drinking from sealed metal bags, enjoying rehydrated vegetables and irradiated meat (2012 was a particularly good year for beef, apparently—they stamped the year on our meat bags, like a fine wine). I wouldn't trade in my fine dining here on Earth, but I wouldn't mind an occasional space food meal every now and then.

MAKING MOVIES

An Entire IMAX Movie Shot in Orbit

Of all the things I did in space—science, rendezvous, spacewalks, etc.—I think the most impactful and lasting was to film the movie *A Beautiful Planet*. It all began one day more than a year before launch, when I checked my iPhone calendar to see where my next class was, and it said Building 9—IMAX. I was intrigued and excited because I'd been watching IMAX movies since I was a kid. In fact, seeing *To Fly!* at the National Air and Space Museum as a nine-year-old was largely responsible for motivating me to become a pilot, and I've seen all the space IMAX movies ever since—*The Dream Is Alive*, *Blue Planet*, *Space Station 3D*, *Hubble 3D*, etc. So when I showed up for my class and found Toni Myers, James Neihouse, and Marsha Ivins waiting for me, I was pumped!

Toni is legendary among documentary directors. If you've ever seen a space IMAX movie, you've seen her work. She's also the recipient of the Order of Canada as well as NASA's Exceptional Public Achievement Medal, two of the highest honors possible. She has simply set the standard for making movies in space; nobody has even come close, and nobody ever will. Sadly, she passed away in 2019 after a brave fight with cancer, and I was honored to have been involved in her final work.

James was Toni's director of photography for her extraordinary run of films, and he, too, is the best. The two of them have trained more than 150 astronauts to be filmmakers, dating back to the 1980s, and their work has inspired hundreds of millions of people around the world. Marsha is a former astronaut and Toni/James protégée herself. She has starred in and helped film several of these IMAX documentaries dating back to the 1990s, and she was a consultant for us on *A Beautiful Planet*, helping to translate between Hollywood-speak and NASA-speak. It's harder than you can imagine.

That first day of IMAX training was awesome. I couldn't believe how lucky I was to have a chance to help film a movie in space. Some astronauts are really into photography, and others not so much. I was definitely in the former camp; I'm the type of dad whose kids are always saying, "Dad, no more pictures!"

Toni laid out our plan: We would spend the next year training on the equipment, learning the basics of cinematography, and getting familiar with the story arc of the film. We would be making *A Beautiful Planet*, a movie about Earth's environmental challenges as well as the cooperation among our international crew. The graduation exercise would be shooting a few scenes in the station simulator and then watching them on a giant IMAX screen. Toni was a great, if humbling, teacher. That I would have a chance to help film an IMAX film during my flight was the best news I'd heard since I was picked to be an astronaut!

Canon partnered with IMAX to provide the equipment, and they gave us some of the best gear imaginable. For still images we used the 1DC professional camera, similar to what is commonly used for professional sporting events, and for video we used a C500. Our original plan was to shoot night shots with the 1DC as time-lapse sequences, because a still camera has much greater light sensitivity than a video camera. We would set the camera to take two frames per second and let it run for a few minutes, taking hundreds of still images. Then the IMAX team on the ground would stitch those still images together to make a video scene. The C500 was a proper Hollywood-quality video camera, complete with myriad accessories to hold it and the monitor and hard drives and brackets together. It was originally intended for daylight shots of Earth, as well as interior shots of astronauts floating. The final camera type was a GoPro Hero, which was lent to us by our Russian crewmates. This miniature camera allowed footage from inside the tiny Soyuz capsule and also from my spacewalks. Most importantly, we had a set of four Zeiss lenses that were absolutely spectacular, enabling many of our first-ever shots from space, such as the night ones.

James, the director of photography, was also our expert instructor, and he set about trying to teach us all the intricacies of this complicated equipment. Setup. Exposure. ISO. Focus. Framing. Focus. Sound. Lighting. Did I mention Focus? One of the things we learned on that trial run on the big screen was that even the smallest error in focus led to a completely useless shot. Focus is digital, and either it's good or it's not, and that was a lesson that I learned time and again. We also had an amazing lesson in sound from the designer of many of the iconic *Star Wars* sounds, Mr. Ben Burtt. I never realized it, but sound is truly half of a movie, and his class motivated me to pay attention to details, like capturing the sound of fans whirring in the background noise of the station, or metal workout equipment banging together like wind chimes, or the sound of air rushing out into the vacuum of space through a hatch. On the space shuttle there was a unique and unforgettable ghost-like moan whenever someone talked on the radio. These sounds are forever etched in my brain, and I hope I captured a few to share with IMAX theatergoers.

> On the space shuttle there was a unique and unforgettable ghost-like moan whenever someone talked on the radio.

Beyond the technical requirements of operating a camera were artistic skills that astronauts aren't known for. But you can't be an excellent filmmaker without artistic effort, so I really followed Toni's guidance. Capture the human element above all else; a dry documentary about geography wasn't our goal. Tell a story, using panning, framing, lighting, and sound. Don't be shy about shooting something as it happens—be as spontaneous as possible, given the limited hard drive space and the fact that it took five minutes to power the camera up. And, of course, always remember to stabilize and focus—a shaky or blurry shot was more useless than a screen door on a space station.

NASA administrators Mike Griffin and Charlie Bolden, as well as associate administrator Bill Gerstenmaier, made all this possible, effectively overriding midlevel managers who didn't want us wasting time filming a movie. In their minds, we had more important work to do. As far as I can remember, I was scheduled for a grand total of one hour of dedicated time for this project. Other than that, every other scene I shot was on my own time, and the same goes for the rest of my crew. Of course, we never skipped scheduled work to do IMAX; it was just such an important project to my crew and me that we were willing to use personal time to make it happen. In case you're wondering, we didn't get paid at all. Though I joke that I got paid twice what my crewmates got paid because I filmed a nude scene (of me showering, above the waist of course). Two times zero.

Once on orbit I tried to work with Toni, James, and Marsha remotely. Toni had a long list of more than 300 shots she wanted: New Zealand, Australia, Beijing, the northern lights, the crew eating dinner together, a cargo ship approaching, etc. This was a good starting point, but frankly it overwhelmed my brain, so she came up with a top ten list. I focused on getting those done while keeping the other 300 in the back of my mind.

James was my technical advisor, but I felt sorry for him. In a normal movie, the director of photography (DP) is intimately involved in every detail of every scene, but for this film he had to trust us. I would call James periodically on our satphone to ask for advice on ISO or lenses or exposure settings. James modified the plan for Earth shots once we were in space; we were to use the 1DC for all Earth views, day or night, and save the C500 for people shots. I requested one exception to this rule. Night lightning was

Practicing photography in a Cupola simulator. Thankfully, the real Cupola in space had glass in its windows, not just a big hole.

spectacular, but the 1DC could only capture it at two frames per second (fps), leading to a bit of a jerky view—the lightning was either on or off. Using the C500 at a much higher frame rate of twelve fps gave a subtler image of lightning flashing. There is a spectacular scene of a giant thunderstorm at night in *A Beautiful Planet*, and it was only possible because of James's innovative DP work.

Marsha helped me track the shots I had taken and manage the memory cards and hard drives that our images were stored on. It was a massive task, one for which we needed a full-time accountant. But alas, we didn't have one, so Marsha and I spent a lot of time tracking those memory cards. Marsha also helped me with shot quality and ideas. During my shuttle flight, she had been the astronaut tasked to help our crew with photography. After a few days, she had reviewed all of our pictures and noticed something. There weren't many pictures of me, because I was usually behind the camera, so she directed my crewmates to take more pics of me until she cried uncle and said, "Enough shots of Virts!" I also had a goofy head lamp constantly floating around my neck, along with a dorky fanny pack around my waist. These accessories were practical, but they made me look like a rookie, so I ditched them based on Marsha's comments. When I got back to Earth and compared photos from those first tacky days in space with the shots after I removed the offending items, I was *so* thankful for Marsha's timely feedback.

Most of the shots that ended up in the film were coordinated with Toni. However, there were a few unanticipated ones, and those were some of my favorites. For example, on Christmas Eve, I was floating in my crew quarters and an idea struck me. I went down to the airlock (where we got into and out of the ISS during a spacewalk) and prepared the set. I put Santa hats on each of the two spacesuits, hung our Christmas tree from the ceiling, and gathered a bag of dehydrated milk and packaged cookies. I taped the cookie bag to the milk to keep them from flying apart from each other and slowly floated them in front of the camera, with "Santa" Sharpied in large letters on his milk bag. We were ready for the big guy to visit our lonely crew of astronauts. Hopefully, his reindeer had spacesuits. That scene was one of my favorites in *ABP*, and it always gets a laugh from the audience.

On another occasion, I was looking out of the Japanese module window and realized just how quickly the outside of the station transitioned from black to white to blue to orange to red to pink as we flew into sunrise. I had an idea: Just before sunrise, I would film the slowly brightening solar panels and exterior of the station with the camera pressed against the window. Then, when the moment of sunrise was imminent, I would push myself away from the window, floating down the center of the module, while it was still dark inside. When the sun abruptly rose, the interior of the station would suddenly be flooded with blinding light. I thought it would be a cool scene, so I gave it a shot. Unfortunately, it was really hard to time the exact moment of sunrise. If I pushed away from the window too soon it would just be a long scene of darkness, and if I pushed away

> Finally, there was a dark green patch of jungle on the horizon as we raced toward it at 17,500 mph, and we came up with an impromptu plan to time the shot.

too late the sun would rise before I had a chance to float into the blackened module. Most of all it was almost impossible to float down the middle of the module holding the camera steady in my hand. It was so easy to push slightly in the wrong direction and bounce off the walls, spinning around. Nonetheless, the first time Toni saw a sample shot via downlink she was so excited because it was an unplanned and unique scene! I never did shoot a perfect one, though, and after many attempts it didn't make the final cut. But getting creative was fun nonetheless.

The most difficult shot I filmed is one that doesn't really stand out in the film. Toni wanted to highlight the Cupola, the module where most astronaut photos are taken. To do this I wanted the camera to slowly float into the module, filming Samantha while she was taking pictures of Earth. There's only one problem—it's really hard to get the right exposure for both an astronaut and the Earth in the same scene, because our planet is so much brighter than the inside of the station. Getting both exposed properly for still photos is much easier because we have a flash, which works for a fraction of a second. But this video clip would be thirty seconds long, much too long for a flash, and the floodlights we had weren't nearly bright enough. So, Samantha got herself awkwardly curled up in the window, camera in hand, gazing off into the distance. Next, we waited for a part of the Earth that was relatively dark. Clouds or water would be *way* too bright and ruin the exposure. The best place was South America, where the Amazon jungle is very dark and, I hoped, would allow the scene to be exposed properly.

Camera ready, lights on, Samantha curled up in a ball on the window, and we waited. And waited. There was seemingly endless cloud cover! Finally, there was a dark green patch of jungle on the horizon as we raced toward it at 17,500 mph, and we came up with an impromptu plan to time the shot. She gave me a countdown as the jungle approached, and I pushed off and slowly floated up toward her, timing the jungle background just right so that both she and the dark green Earth background would be exposed. As soon as I had thirty seconds of good video, the jungle was again covered with überbright clouds, badly overexposing the camera. We had just barely squeezed that shot in. It made it into the final cut, and though you probably wouldn't notice it when you see *A Beautiful Planet*, every time I see it I feel like a proud parent!

Though there are a lot of important aspects to the space station—the engineering achievements, the science, and most of all the international cooperation—I still believe that the most important work that I did while in space was helping to make *A Beautiful Planet*. During the grand opening at the Air and Space Museum in Washington, DC, its director told me that over the next decade more than a million people would see this movie. That was a shocking statement. What a far-reaching impact we had, inspiring people young and old and hopefully making them wonder what is possible in the future.

Thanks to the lifelong dedication, creativity, and leadership of Toni Myers and James Neihouse and the whole IMAX team, this movie was possible. Most of all, thank you, Toni, for being such a wonderful person; you are loved and missed.

ZZZZZZZZZZ

Sleeping While Floating Is Awesome

There are a few things that everyone is apprehensive about before their first spaceflight. Oddly, death was not one of my concerns. My biggest worry was messing up. You've heard of the Lord's Prayer; well, during Project Mercury, there was a prayer attributed to America's first astronaut Alan Shepard, "Shepard's Prayer." *Dear Lord, please don't let me f . . . up.* I'd be willing to bet that the fear of publicly making a mistake is high on the list of every astronaut's concerns, rookie or veteran. For me there were other worries—how would I adapt to floating, would I get sick, how would I get along with my crewmates, could I get all of my work done? And would I be able to sleep in space?

The short answer to this last question is a resounding yes! My first flight was a two-week shuttle mission to finish the construction phase of the space station, and boy, were we busy. Houston scheduled our lives in five-minute increments for each of our fifteen days in space. When I finally made it back to my house after landing, I was so exhausted that I slept for thirteen and a half hours straight, something I hadn't done since I was six months old.

My first night in space came at the end of a very long work day; we had woken up about ten hours before launch and then spent the whole day after launch converting the shuttle from rocket to spaceship. I had a headache and felt dizzy and had about a million things to do filling my brain, all while adapting to the alien environment of space. By bedtime I was wiped out, but also disoriented. The feeling of weightlessness with no up or down was a little overwhelming, I couldn't keep track of stuff because everything was floating, and on top of it all there were six of us on the shuttle middeck. Imagine seeing zipped-up sleeping bags covering all the walls and ceiling and floor of a volume equivalent to a large bathroom or walk-in closet—it was cool and disorienting all at once, like a scene from *Inception*.

Before falling asleep, I gathered what I would need for the next day. I put tomorrow's underwear and clothes in my sleeping bag. I also got my blindfold (the sun rises every ninety minutes in orbit, so you need to cover your eyes), ear plugs (yes, people snore in space), and iPod to listen to some music, and I also kept my flashlight available, in case a middle-of-the-night bathroom break was required. Next, I found a place to sleep, just like a camping trip, but in space the walls and ceiling are also available. I usually picked out the starboard wall of *Endeavour* during STS-130, or the ceiling of the Columbus module when we were docked to the ISS, clipping the top and bottom of my sleeping bag to the wall.

The act of falling asleep was what I had been concerned about. My fatigue overcame my disorientation. I put on some music, closed my eyes, thought about the day's amazing events and what was coming tomorrow, and wondered about Earth . . . zzzzz. . . . The next thing I knew, wake-up music from mission control was blaring on the speaker and it was time to wake up. I usually fell asleep very quickly and slept hard through the night.

Wake-up music was a tradition for several decades during the shuttle program; crews were awakened every morning with a song chosen by their families or flight controllers. Some of those songs were pretty funny, some emotional, and some made you go "Huh?" But it was a nice tradition. I had three songs for me on STS-130: "Give Me Your Eyes" by Brandon Heath, "In Wonder" by the Newsboys, and "I'm Gonna Be (500 miles)" by Steven Curtis Chapman. On our seventh day in space, in the middle of installing and unpacking thousands of pounds of gear and equipment, my crewmate "Stevie-Ray" Robinson had the song "Too Much Stuff" by Delbert McClinton playing, and boy was that appropriate. Kay Hire had "Window on the World" by Jimmy Buffett, a song he wrote for our mission, because we installed the Cupola, the amazing seven-windowed module that is the coolest thing ever flown in space.

My first few nights in space I took Phenergan, a motion-sickness medicine that also helps you sleep, but I didn't need it for either reason. At bedtime

I was out like a baby, though my shuttle and station flights had one thing in common: I didn't get enough sleep. On the 200-day mission I would use Sundays to catch up, turning my alarm off and sometimes sleeping until noon. I figured if my body needed sleep, it would sleep, and that seemed to work. But the other six days per week, my Omega X-33 watch would faithfully beep every morning. Station crews didn't get wake-up music, thankfully.

During my long-duration mission, we each had our own crew quarters, which was a private cabin about the size of a phone booth. Not only was having private space critical for our psychological well-being, it was also great for sleeping. I did not clip my sleeping bag to the wall; I simply free-floated in the small cabin. I'd put my entire body in the sleeping bag, head and arms and all. The airflow from the cabin vents would slowly spin me around so that every morning I awoke in the same corner. It was kind of funny, like being in trouble with the teacher when you're seven years old and having to stand in the corner. The feeling of floating and being completely sensory deprived, eyes closed, no sound, is stunning and surreal. It really feels like you are in a void, with nothing else in the universe other than your being; everything simply fades to black. My mind has never been so free of clutter. A friend of mine described having the same sensation while cave diving in a giant black underwater room. I think it's a similar experience, but floating in space in complete blackness was sublime.

One of the best things to do at nighttime was to put on my Bose headset and listen to music as I drifted off to sleep. I spent a month listening to Hans Zimmer and his *Interstellar* soundtrack. After about four months in space, our Russian colleagues received some "sounds of Earth" MP3 files from their psychologist, and the whole crew absolutely loved them. Rain, waves, jungle sounds, crowded café. These sounds connected me to our planet more than I imagined, and I ended up drifting off to sleep to the sound of rain for a month.

One question that I'm routinely asked is "Did you dream in space?" I often dreamed of Earth during that month when I fell asleep to the sound of rain. Bizarre dreams, in places and with people whom I hadn't seen in years. I remember one particularly detailed and scary one, climbing through abandoned houses and forts through a forest on Earth, with someone or something chasing me. It was jarring to be so vividly on Earth in a dream and then wake up floating, take my earplugs out and blindfold off, and hear the continuous hum of the fans and the artificial glare of cabin lights. Dreaming while weightless was one of the most sublime experiences of my life, if sometimes scary. It was even more powerful than I had imagined before I first left Earth.

Interestingly, when I wasn't listening to rain or other sounds of Earth, my dreams were about space. Of me floating through dark asteroid fields, no spaceship, just moving through the blackness of space trying to avoid the gray asteroids. I wondered if I was subconsciously thinking of the dangers of space debris impacting the station. They weren't nightmares, but I think they captured the essence of space. Though we usually think of planets and galaxies and nebulae when we think of space, the reality is that the vast majority is just dark, cold blackness. And my dreams often took me there.

NO SHOWERS FOR 200 DAYS? NO PROBLEM!

Bathing in Space

The space station is, in many ways, a thirteen-year-old boy's dream. You can float around like Superman. You can eat whatever you want and your parents aren't there to nag you. You have your own room and you close the door and nobody tells you to clean it. Best of all—*no showers*! For 200 days in a row!

Now for those of you who *aren't* thirteen-year-old boys, you may be saying to yourself, "Gross!" So let me ease your fears. It is possible to take care of hygiene, the NASA term for washing up, pretty effectively while in space. There are several aspects to staying clean, each with very different solutions: showering, brushing your teeth, washing your hair, cleaning your clothes, etc. Let's start with the big one—showering.

NASA actually flew a space shower on the Skylab space station in the 1970s. It was about the size of an Earth shower, with a big bag around it to capture flying water droplets. I think it worked OK for the astronauts' weekly shower, but the real problem was the cost. Today, a half-liter water bottle costs about $20K to send to the ISS. You can imagine how many of those water bottles are required to take a shower. Even though much of that water could be recycled, a shower is still a massive and expensive thing. So we found a better way to solve the problem of stink.

I used a wet towel to wash every day during my seven months in space. It worked great, and I never missed taking showers at all. The mechanics were straightforward. I would start out by filling an empty drink bag with hot water, which I squirted into a towel. NASA had budgeted for each astronaut to get a new towel every two weeks, so there were plenty available. I would get naked and wash from head to toe. I would also use rinseless shampoo to wash my hair, the same kind that hospitals use for bedridden patients. When done, I'd use a separate towel to dry off. And voilà, clean astronaut. The whole process took maybe fifteen to twenty minutes, but it got me clean and not stinky, which my crewmates appreciated. As a backup, there was deodorant.

I usually washed after working out. Normal activity in space doesn't get you dirty or sweaty at all—a day without exercise was a day when you basically were as clean at night as you had been when you woke up. And your hair didn't get messed up because you didn't sleep on a pillow or go out on a windy day. But exercise was a different matter altogether: a good thirty-minute run on the treadmill or hour-long exercise on the ARED weight-lifting machine made me sweat. A lot. And, therefore, need a "shower."

So floating there naked for ten minutes as you washed off would be, to say the least, awkward. Even though we are all brothers and sisters, that situation is best avoided.

The location where we washed varied. The ground engineers wanted us to clean up in Node 3, the life-support module where the bathroom, ARED, and treadmill are and the Cupola is attached. It seemed to make sense that it would also be the hygiene location. However, Node 3 is the busiest module on the ISS; there is always someone working out, running on the treadmill, going to the bathroom, looking out the Cupola, performing maintenance on some critical ISS life-support equipment, etc. So floating there naked for ten minutes as you washed off would be, to say the least, awkward. Even though we are all brothers and sisters, that situation is best avoided. Nobody needs to see your crewmates' junk floating weightlessly as they wash while you run on the treadmill.

So my crew used a module called PMM for hygiene. PMM is a large stowage module that is both the garage and closet of the ISS. It's full of equipment and usually doesn't have any astronauts working in there, so we put up a curtain to cover the hatch opening and agreed among ourselves—wash up in the PMM, close the curtain while you're naked, and open it when you're done. That worked. However, while you're wiping down with a wet towel or washing hair, water droplets inevitably float off in all directions. If you have seen the IMAX film *A Beautiful Planet*, you've seen my shower scene, in which water droplets fly everywhere in 3D. The other agreement we had was that we'd clean up any mess we made so that the equipment in PMM didn't get too wet or moldy. Also, we each commandeered a location in PMM for storing our shower gear—towels, shampoo, soap, deodorant, etc.

There were a few days when the PMM wasn't available because it was being relocated from a hatch on Node 1 to a new port on Node 3, so we had to find an alternate location. For me, the JLP was the best alternative. JLP stands for "small storage module attached to the Japanese lab." It is out of the way and usually free of astronauts, so it's a good place to wash. The Japanese flight controllers weren't happy about us using their module for hygiene because they wanted to keep it pristine. But the crew needs to clean up somewhere, so we used it, but sparingly. The Russians used a module called FGB for hygiene—their storage module. It was the first segment of the station launched, back in 1998, and it's in the middle of the ISS. They would partially close the hatch when washing, but if you had to float through there while they were showering, it was either wait until they were finished or have a very awkward moment as you floated by, avoiding unintentional contact. But hey, sacrifices must be made in the conquest of space.

A few of my colleagues talked about showering in their own sleep station, but in my mind that was a bad idea. I didn't want to get my sleeping bag, laptops, and personal gear wet and moldy. I don't think this happened very often.

In addition to using hot water and normal towels, we also had towels that were designed for campers and were pretreated with soap. I would take one, fill it with hot water, wait a few minutes, and voilà—hot, soapy towel with which to shower. Which was great. For one shower. Then the next day the soap was mostly gone, so the towel itself could be used again, but pretty much with no soap. Also, there were two types of these camping towels. One was bad—it was basically a paper towel that would disintegrate into a million crumbs as you washed yourself. Our crew had an endless supply of these, and I tried my best to use them up so that future crews wouldn't have to deal with them. There was also a much better towel, more substantial, that you could

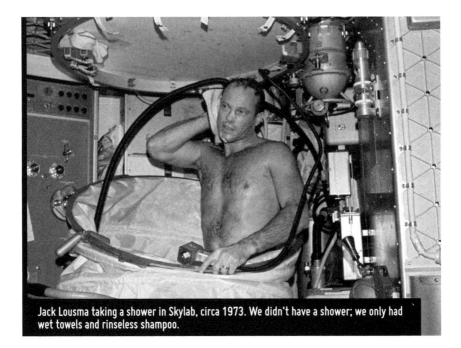

Jack Lousma taking a shower in Skylab, circa 1973. We didn't have a shower; we only had wet towels and rinseless shampoo.

use for several days. We used these camping towels for getting wet and soapy and a normal bath towel for drying off.

The soap bags were exactly like drink bags, except you got soapy water instead of tea or coffee or Tang when you filled them full of hot water. I used one of those every other day, until after two weeks I noticed that my supply was out. I called Houston and asked where my next stash of soap was, and they sheepishly answered, "Uh, well, that's all the soap you have for the whole mission." I had misread the usage spreadsheet that told me how often I got a new pair of underwear, deodorant, T-shirt, and, you guessed it, soap bag. It was once every two *weeks*, not two *days*. Oops. I found a few emergency soap bags stashed here and there, but for the most part the only soap I had during the 200-day mission was from the camping towels. At least I didn't stink, though the lack of soap was bad for my skin.

Another major daily hygiene routine is brushing teeth. Every astronaut is given a few toothbrushes and tubes of toothpaste to last their half-year mission, and I've never heard of anyone running out. There are basically two techniques for brushing your teeth. Some folks spit out after they are done, and some swallow the used toothpaste. For me, taking the extra time and using up one of my limited supply of towels to spit into was more hassle than it was worth, so I learned to brush my teeth with just a little bit of toothpaste and wash it down with a little water. It wasn't so bad, and I got a few months out of one tube of Crest, brushing twice a day.

An interesting aspect of washing in space is dealing with wet towels. They are left out to dry, one end clipped to a wall or ceiling. The whole towel stands straight up, as if it were hanging from a rod here on Earth, so you have to find an out-of-the-way location for it to stick out into. It dries out within a few hours later, from the cabin air blowing over it, which is good for two reasons. First, you can use your towel again. The (good) camping towels could usually be used for a few days, and a normal body towel could last for several weeks. The second and most important reason for drying these towels is that the water is recycled. It goes from towel to ISS atmosphere, where it is reclaimed as humidity and then purified into drinking water or split into H_2 and O_2 for breathable oxygen. The H_2 is combined with reclaimed CO_2 to make water, which is used to drink or rehydrate food, and so on and so on.

Overall, astronauts keep clean, sort of. During my time on the ISS, and comparing notes with many fellow astronauts over the years, skin problems were common. Rashes, bumps, red spots, and general discomfort. Nothing too serious, but bad skin is a very common ailment for ISS astronauts. There are several potential reasons for this. Perhaps certain types of mold that are on board irritate skin, perhaps there are adverse effects of weightlessness and fluid shift on skin, or maybe having one bag of soap isn't enough to keep skin clean. One of my top recommendations after Expedition 42/43 was to improve the skin cleanliness situation on the ISS. Although we didn't stink, we needed better skin care in order to prevent some of those rashes.

So, again, if you're a thirteen-year-old boy, a flight to the ISS is perfect for you. No showers for half a year or more! But you have to swallow your toothpaste and have nasty skin rashes.

THE GLAMOUR OF SPACE TRAVEL

Going to the Bathroom in Space, *Uncensored*

O K. Everyone knew this chapter was coming. There are certain questions that astronauts are always asked: "What's it like in space?" "Is it hard to adjust to gravity after you come home?" "Did you see aliens?" But the most common question, no matter what the age or country of the person asking, is "How do you go to the bathroom in space?" The answer is simple. Very carefully. Seriously. This is one procedure that nobody wants to mess up.

The actual mechanics are pretty straightforward. On Earth we have gravity to make sure everything goes in the right direction. In space we use airflow. Fans blow air through both the urine hose as well as the number two can to help everything get to where it's going, and then stay there. In fact, there was an emergency procedure on the ISS to quickly close the valve on the hose in the event of a fan malfunction—because waste would begin to float out. And that would be a bad day in space. Imagine a Porta Potty at your local football stadium being tipped on its side and then floating. You just don't want to go there.

As the shuttle pilot, one of my jobs was to keep the WHC (NASA acronym for toilet) in working order. As a rookie I was very motivated to do a great job, so I went all in, determined to keep the cleanest bathroom in shuttle program history. Before launch, my first task was to select cleaning supplies—toilet paper, baby wipes, and disinfectant wipes. When we saw the actual bag of supplies I had ordered, we laughed out loud. It was huge! There was enough to run an average household's facilities for a year, but I wasn't fazed; I was determined to keep that pot clean. After our two-week mission we had barely made a dent in all of it, so I left a huge bag of supplies for the station crew after we undocked. When I came back to the ISS in

2014, there were probably still packs of Huggies wipes that I had left there in 2010.

A few words about space potties. The shuttle had the WHC, and the ISS has two Russian-built ACYs (Russian acronym for toilet), but the basic principle behind both of these devices is the same. There is a hose for urine and a big can for poop. On our shuttle mission, each crewmember had their own funnel for the urine hose. There were different shapes for men and women; I never asked why and don't really want to know. But on the ISS we all shared the same funnel, men and women. The shuttle brought all of its waste back to Earth (bless the ground crew who had to clean it out), but the ISS ACY had disposable KTO tanks (Russian acronym for poop bucket) that would be packed into Progress cargo ships to eventually burn up in the atmosphere, in a streaking poop meteor hurtling five miles per second over the Pacific Ocean to a fiery death.

The big difference between the US and Russian systems is how they handle urine. The shuttle simply brought it back to Earth. The Russian toilet stored it in a ЕДУ (Russian acronym for pee bucket) that could also be destroyed when the Progress burned up, but it was more often transferred to the American segment for recycling. That's right, we turn yesterday's coffee into tomorrow's coffee by recycling urine. This is an extremely useful capability because it costs about $40,000 to launch a liter of water into orbit, and recycling as much water as possible saves taxpayers serious money. Not to mention that when we eventually go back to the Moon and on to Mars there won't be much, if any, water there for us to use, so we will need to recycle. As far as the "yuck factor" goes, it wasn't even an issue; the water on the ISS tasted completely normal. You just had to close your eyes and get over the yellow color—just kidding. It looked and tasted and smelled just like normal water; the US recycling system really was remarkable.

The first time doing anything in space was always a little nerve-racking, and when it came to using the bathroom for the first time it was especially so. I absolutely did not want to be the guy who broke the toilet, so I very carefully followed every step of the checklist. Yes, there's a NASA procedure for number one and a different one for number two, and both were fairly time-consuming. After a few days I got pretty efficient, but I still diligently followed that checklist, step-by-step, until one fateful time on flight day eleven.

By my eleventh day in space I had gotten pretty cocky. I wasn't half-bad at floating, could keep track of stuff without losing it, and could get dressed in the morning without banging off the walls. When it was time to take a bathroom break that afternoon, I was in a hurry to get back to my work, and I decided to go number two without using the checklist. It's important to understand how the poop bucket works on the shuttle—it was a big tank, the size of a small dishwasher, that you sat on top of. First you had to open what we called the guillotine, an ominous-looking hatch, and when you closed it you had to make sure there was no "debris" in the way, or you'd have a mess on your hands. Once the guillotine was open, you sat on it and placed handles over your thighs to prevent you from floating away.

It was *extremely* important to have proper aim—you had to deliver the package on target, and you did not want it to scrape against the side of the guillotine or the seat, or you'd have another mess. Believe it or not, the ground training potty at the Johnson Space Center in Houston had a camera down in the hole, and new astronauts would sit on it and practice aligning until they had the muscle memory of what proper aim felt like. Having good aim was that important.

The first few times I went through this procedure in space, it was difficult. There was no gravity to help the poop come out, so I had to provide all of that force with my muscles, which involved a lot of straining and grunting. At times I thought I'd pop a vein in my forehead. This procedure was similar to the AGSM (Air Force acronym for anti-g straining maneuver, used by fighter pilots when pulling g's). Thankfully, I was always able to successfully complete the mission; some folks on other missions couldn't and needed help with medication.

On that fateful STS-130 flight day eleven, I was in a hurry. I floated into the WHC, closed the small curtain, got myself aligned, and started to grunt. Everything was working fine until something didn't feel quite right. I stopped immediately and remembered—*ugh*—I had forgotten to open the guillotine. OMG, that was not what I wanted to happen. So I slowly floated up a few inches, looked down, and yup, there it was, gently swaying in the breeze, very lightly attached to the closed guillotine. Without panicking, I quickly got some tissue, wrangled the offender, floated it next to me as I quickly wiped up the minor mess, opened the guillotine, floated it down into the bucket, and then finished my business, this time with the hatch properly open. Disaster narrowly averted. My crewmates and I still laugh so hard about that incident—the perils of spaceflight.

> When people eventually fly to Mars, something as simple as a broken toilet could end up killing the crew, so it's a very important subject—even if it does lend itself to thirteen-year-old-boy humor.

Besides following the checklist and keeping everything clean, there was toilet maintenance to be done on the shuttle. First, midway through the mission I compacted the solid waste. The big toilet bucket had a net inside, and I rotated it to scoop up everything and compact it, freeing as much room as possible for the second half of the mission. Before compaction, looking down into that bucket reminded me of the asteroid scene in *The Empire Strikes Back*. Some colleagues described it as a "turd-nado." Whatever the term, after a week of six people using the facility, the toilet needed compacting to make room for the next week.

This was done using a torque wrench, a socket wrench that limits how much force can be applied to a bolt. I used it to slowly turn a bolt that moved the net until it had swept up all of the debris, while staring at the torque wrench very, very, very nervously. Because if the torque limit was reached before the net was finished sweeping the whole bucket, I would have to get in the toilet and do it manually. NASA didn't want us putting too much stress on that fragile bolt, ergo the protection from the torque wrench. Several of my crewmates watched me as I performed this procedure, and when the net was fully compacted without reaching the torque limit we let out a big cheer. Especially me. On an earlier shuttle flight, one of my fellow astronauts wasn't so lucky—the torque wrench had triggered, stopping the compaction. He then had to put on elbow-length gloves, the kind large-animal veterinarians use, grab the poop spatula (yes, it's a real thing), and stick his arm down into the asteroid field, herding everything to one side so the net could be swept across the bucket. He finally had all of the asteroids compacted, with the toilet ready to accept more.

Staring at the torque wrench during WHC compaction, sweat beading on my forehead, was one of the scariest moments of my time in space.

The Russian system was different than the shuttle's; it used a disposable KTO bucket that was the size of a small kitchen trash can. You just had to look inside and see when it was full. There was no net to compact, so when it started to get full you would take a gloved hand and push things to the bottom. The good news with this system is that each time you pooped you went in a small bag, so the asteroids in the KTO were just bags floating around, like Ziplocs of pudding, so it wasn't too terrible to compact them. I was proud of a system I invented to indicate remaining capacity. When a new KTO was installed, I drew a smiley face on it with a Sharpie. A few days later, I drew a not-so-happy face next to the original. Around the ten-day point, I would draw a frowny face, and then on day eleven, twelve, or thirteen, it would get a face with Xs for eyes, showing it was dead, and it would be replaced with a fresh, empty KTO. We were supposed to go eleven days for our US segment crew of three, but it was a matter of pride to extend it to twelve, thirteen, or even fourteen days. Of course, this depended on the crew—three offensive linemen would probably not have been able to go that long. But, luckily, we were a little more average-size.

This is definitely a humorous subject, but it's also a very human one. Using the bathroom is something we all have in common, and in space it's something that requires special equipment and procedures to perform successfully. The technology to recycle water on the ISS is truly remarkable, and it has even been used to make clean water available to remote communities around the world. When people eventually fly to Mars, something as simple as a broken toilet could end up killing the crew, so it's a very important subject— even if it does lend itself to thirteen-year-old-boy humor. But that's OK, I'm a fighter pilot. I was proud to boldly go where (few) had gone before. And I still stand by the assertion that my toilet—the STS-130 PLT—was the best ever.

SATURDAY CLEANING

An Astronaut's Work Is Never Done

Most of our daily routines have a few things in common, no matter what country you are from. We shower in the morning. We go grocery shopping after work. And we clean on Saturday mornings. Vacuuming, picking up, dusting, cleaning the kitchen, yardwork—it's what Saturdays seem to be made for. And it's not any different on board the ISS. Although most people think being an astronaut is a glamorous and nonstop life of launching, doing spacewalks, and being a hero for kids while wearing the blue jacket, the truth is that life in space can often be, well, life.

Several types of messes happen in space, each requiring its own unique cleaning procedure. The first is general clutter, like what happens on your kitchen counter or bedroom nightstand. Stuff accumulates and needs to be put away. In space, things tend to accumulate on the handrails. These are metal bars strategically placed throughout the ISS on floors, ceilings, walls, and especially the junctions between modules. We move around by grabbing these things and pulling ourselves to the next location, and once we get there we stabilize ourselves by wedging our feet under them. Floating in weightlessness requires a lot of adaptation, and handrails are one of the best inventions of the space age to make astronaut lives more convenient.

Beyond helping astronauts move around, handrails are also a very convenient spot to leave things. Much of our equipment is held in place by brackets or Velcro or duct tape, and it is very easy to stick these things onto a handrail and forget about them, leaving them there. Camera brackets were a significant offender; they were often clamped to the middle of handrails, right where an astronaut flying by would reach out to grab. Imagine yourself floating down the center of a module when you get to the end. You need to reach out and grab the handrail to redirect your motion by 90 degrees to float into the adjacent module. But when you reach out, it's covered with big metal camera brackets. You have to quickly reach down and grab a free spot to swing around before crashing into the end of the module. Disaster avoided, but it was annoying. My OCD was not a fan of handrail clutter, so I was always spending time cleaning them up. It was the space equivalent of Lego pieces left on the floor by the bed.

Another type of mess is from food. At home, your food mess usually ends up on the floor, the kitchen counter, and the stove—all predictable locations. But in space, soup droplets or chicken parts literally become their own free-flying satellites, taking random paths to land on hard-to-find locations. Often airflow from the ISS fans captures the free-flyers and sucks them into a filter, where we find the most dirt. But in Node 1, the location of the American kitchen, there are food particles and stains everywhere. That module has been in space for more than twenty years now, and it's, shall we say, a little ripe. So a big part of Saturday morning chores involves taking disinfectant wipes and floating around looking for stray food stains to wipe up. In the science modules, like the European and Japanese labs, there is almost no dirt. But in Node 1, where we eat, and in Node 3, where we have the bathroom and clean up, there are always spots that need to be wiped up.

Vacuuming is a big part of Saturday morning chores. The first challenge is to find a free electrical outlet—almost all of the ISS electrical plugs are taken, and there is an official technical diagram, called the plug-in plan, designed by a team of NASA engineers that specifies which outlets are allowed for the vacuum and which aren't. Once you find a free outlet, you plug in a very long extension cord to drag the vacuum between modules. Running a vacuum in space is something that every astronaut has played with at one time or another, because as the fan spins inside the vacuum, there is nothing to resist its angular momentum, so it wants to spin violently in the other direction. Kind of like a balloon that flies around randomly when you let it loose. I know the physicists are cringing about the scientific inaccuracy of this description, but it's kinda like that. Whatever the technical reason for the propensity to flail around, we secure vacuum cleaners with a strong Velcro strap to a handrail. Then you get to work, floating around the ISS looking for vents and filters, where almost all of the dirt is.

Most of these filters are pretty easy to get to, but some require coordinating with Houston to turn off the ISS fans while the filter is removed for vacuuming, because you don't want them sucking down big pieces of debris with no filter to protect the fan. Then you vacuum dirt off the filter, which is a lot like taking dryer lint out of the filter on Earth. I honestly never wanted to know what was in the filters—I just cleaned them off and was done. One time during a training session, unfortunately, I asked my "crew systems" instructor and they told me. It's part lint from clothes and fabric, part dust, part food particles, and part skin.

That's right, skin. Our epidermis is constantly renewing itself and therefore shedding, and in space it's no different. Except on Earth it falls to the ground and just gets vacuumed up or eaten by the dog or whatever. In space it floats around, hopefully to be captured by one of these filters. What's worse, in space we don't use our feet very much, so the calluses on the bottom of your feet begin to shed and disappear. A fellow astronaut once made a video of himself taking off his socks and scratching his feet while in space, and there was a giant cloud of foot skin flakes. Disgusting. Hopefully, the fans and filters suck it up to keep us from breathing it in.

The vast majority of astronauts realize that their mother isn't there and that everyone has to pitch in. During my time on the ISS, we had a system of rotating which modules you were assigned to clean, and everyone had two each Saturday. On weeks when you had Node 1 (eating) or Node 3 (exercise and bathroom), you knew you had some work to do. On Saturdays when it was your turn to clean Columbus and JEM (the European and Japanese science modules), you could high-five yourself because you'd be done quickly! The Russian cosmonauts cleaned up the Russian segment using the same techniques. This system of everyone pitching in and getting it done worked very well. It's not the most glamorous part of the job, but keeping our home clean was probably one of the most important things we did.

WHERE OVER THE WORLD ARE WE?

Recognizing Places on Your Planet

On my first spaceflight, STS-130, we installed the Cupola. As I've mentioned, this seven-windowed module is an amazing place, giving astronauts the best imaginable view of our planet and the universe— it is truly spectacular. I had the privilege of opening the window cover for the first time and taking in the intensely beautiful light from our planet; it was a scene that took my breath away.

After taking a few moments to enjoy the spectacular view, we continued the task at hand, making sure each window cover would open and close, verifying that the equipment inside the module worked, plugging in electrical cables, removing brackets, assembling vacuum fittings and cooling lines, etc. A few hours later, when the initial rush of work was complete, someone had the idea, "Hey, why don't we open up all of the window covers and give ourselves a nice view while we work?" So we did, and it was absolutely awesome, seeing our planet floating by below while we worked diligently in Node 3, adjacent to the Cupola.

About an hour later, I was busy removing launch bolts (designed to hold equipment firmly in place during the shaking and vibrating of launch) when all of a sudden the entire module was bathed in a pink-red glow. It was disconcerting, to say the least. I quickly floated down to the Cupola and looked outside—and down below was the Australian Outback, drifting by slowly in the window. Spectacular! The bright and intense colors of western Australia, red from the iron in the soil, had completely lit up and changed the vibe inside our little spaceship. I was blown away. And would have that experience repeated many more times as I orbited our planet another 3,000 times.

I never expected to know countries and regions by color, but that's exactly what happened. Instead of knowing France by the Eiffel Tower or by its food, I began to know it as a green and often cloudy place. Instead of knowing East Asia by its food and bustling cities, I began to know it as a place covered in bright lights at night and smog during the day. Instead of knowing the Middle East as a land of exotic culture and scorching heat, I knew it as a vast sea of beige, pink, and red deserts, stretching from one edge of the planet to the other. Instead of imagining central Africa as a place of tribal villages and wild creatures, I came to know it as the deepest green, a vast, dark jungle usually covered by thunderstorms.

Flooding. A one-word description of light as it pours into the spaceship and into your eyes and brain and soul. From the planet, from the sun, from the moon and the stars, there is a never-ending flood of light out there—all you have to do is take the time to look and notice. There was always something interesting, even spectacular, every time you looked out a window. Though some of my colleagues confided that after a while looking at the planet got repetitious or boring, and that they would prefer to do work inside the station than look outside, I treasured every glance I stole out those precious Cupola windows.

Oceans. If you were randomly deposited in orbit, the odds are that you would be over an ocean. They are almost everywhere, and even if you're not directly over one, you can usually look in the distance and see one. But they are not all the same. I found their blue colors to be different. The Atlantic looked like a cold blue to me. The Caribbean was full of large swaths of warm turquoise and aqua blue—some of the most beautiful colors and views I've ever seen. Intense and stunning, words just can't describe them. The Pacific is, well, huge. Gigantic, actually. We would be flying along for thirty minutes, 17,500 mph, 8 kilometers every *second*, and be over the Pacific the whole time. In the South Pacific, there is some of that Caribbean aqua blue color in the many atolls that dot the ocean, especially in Australia's Great Barrier Reef, though it's honestly nothing compared to the Bahamas and Cuba. There are always thunderstorms in this part of the Pacific, and puffy cumulus clouds floating below. In the far North Pacific, there were usually extreme cloud formations, scalloped low-altitude clouds that went on for hundreds of miles. And when there was a low-pressure formation, you could see massive hurricane-like storms, with the swirling patterns resembling galaxies whose arms stretched out for hundreds of miles. These storms were not solid cloud formations the way actual hurricanes were, but I imagined that the weather was pretty bad down there. And I now understood why *Deadliest Catch* always seemed to feature huge waves and bad weather near Alaska.

Speaking of bad weather, there is nowhere on Earth like the Southern Ocean, a newly named body of water that is to the south of Australia, South America, and Africa. The clouds there did not look like clouds anywhere else on Earth; they were often angry-looking, swirling storms, similar to those in the North Pacific, though they swirled clockwise instead of counterclockwise. Actually, there were almost always clouds there—it was an ocean that did not look fun to cross on a boat. I was happy to be in space when over that part of Earth. The Indian Ocean seemed to me to be a little more intense in its blue color, and I remember seeing that turquoise/aqua color in the Maldives at night, lit by moonlight, dotted by a few sparse city lights.

There were three locations on Earth that always seemed to have thunderstorms—the Amazon, central Africa, and the South Pacific. If we were passing over these regions at night and I had thirty spare minutes, I would go down to the Cupola, turn off all the interior lights, let my eyes adjust, and float there while watching the most spectacular light show. I still remember seeing lightning while over South America at dusk; I could see lightning bolts flashing while still seeing the details of the clouds in the remaining daylight. Words can't do that justice. The first lightning that I saw from space was actually over the American Southwest, near New Mexico. There were red flashes, which was bizarre, the only time I saw that color associated with lightning. I'm still not sure why that was—perhaps it was a dust storm being illuminated. Whatever the cause, it was beautiful.

Hurricanes, cyclones, typhoons, and tropical storms are part of life on our planet, and, if you spend a few months in space, that fact will really sink in. I saw twenty-three different tropical systems during my 200-day mission, and they were all unique, spectacular, and frightening. Though they were amazing to see, I always kept in my mind the fact that those monsters were potentially dangerous and deadly and would be causing misery down on my planet. The most powerful one I saw was Maysak, a Pacific typhoon. It was, in fact, the strongest storm ever in the month of April. Our whole crew gasped when we saw it for the first time because the eye was so huge, well defined, dark, and ominous. It looked like a Hollywood special effect, except it was real. Most hurricanes had ill-defined eyes, spread out or partially broken, but Maysak was spinning and churning, as perfectly defined as if it had been going to the gym. I felt for the people in the Philippines who had to endure its wrath.

Earth's landmasses are pretty spectacular as well. While the red Australian Outback left a first and lasting impression, flooding the interior of the ISS with its color, it is not the only colorful desert. The Sahara and northern Africa are also vast, a little more pink and less red than Australia, but equally colorful. I took hundreds if not thousands of pictures when flying over this part of the world. The combination of deep blue sky, ocean, and endless red/pink/orange sand was mesmerizing. The northwestern part of the Sahara, Earth's largest desert, around Algeria, is a dark orange with black mountains. Just spectacular. Across the Red Sea we flew over the Saudi peninsula, where there are equally spectacular and colorful deserts. In the southern part of Africa there was the Namib desert, and one day when flying over it, I really noticed how big the sand dunes were. Those were visible in many places, especially Saudi Arabia, but the dunes in Namibia looked big. It turns out that they're the biggest on Earth—some are more than 2,000 feet tall. It impressed me that I could see them that clearly from space.

One of the strangest places on Earth was the Gobi desert and the area in the western part of China and Mongolia, stretching down to the Himalayas. What struck me most was how cold it looked. You can't see temperature by looking out a window, but if you could, that place would register as freezing. Another thing that stood out were the man-made structures. There were roads that were perfectly straight for hundreds of miles, and others that made intricate patterns and different geometric shapes—either way, it was clear that people had been busy down there. I'm not sure if they were for agriculture, science, or most likely military purposes, but they sure were visible from space. As that desert crept southward into the Himalayas, there were a lot of small blue lakes, and I always thought how strange it would be to have a lake in the middle of a desert. In reality it probably wasn't true desert but more of a rocky country, but I wondered what people did there. Fish? Vacation? Or were there even any people there at all? Another thing stood out to me—the Himalayas aren't that big. Not from the vantage point of orbit. You go from cold desert in the north to a snow-covered line of mountains to the green jungles of India pretty quickly.

The Americas have a very large desert region between the American Southwest and Mexico, but from space it never reminded me of a true desert, certainly not like the Sahara or Saudi peninsula. There were so many mountains that it seemed more like a rocky place than desert. But having lived in Arizona, I can say it really is a desert minus the sand dunes. Farther down that landmass you come to South America, and flying from west to east, the Pacific turns into the Andes turns into desert turns into green and the jungle of the Amazon very quickly. The Altiplano of Bolivia is home to some very spectacularly colored terrain, blues and reds dotting the compact beige that dominates the region. I don't think anywhere else on the planet changes terrain types as drastically or quickly as that western corner of South America.

Rivers are a fascinating part of our landscape, but you usually can't see them, especially in forested areas. The Amazon jungle is one area where this is especially true; there are hundreds of rivers, some thousands of miles long, normally hidden from view. Unless the sun reflected off them, and then there were a seemingly infinite number of rivers down there. The Congo was similar; the jungle in that part of Africa is dark, and the green was so dark it almost looked black. Also no rivers—until the sun reflected the majesty and length of that great river system.

Nowhere was this effect more prominent for me than over the South Pacific island of Papua New Guinea. That big island is covered in jungle, and one day I flew overhead at orbital noon, with the sun directly above me. The island went from a mountain jutting out of the Pacific, covered by a dark green jungle, to an explosion of hundreds of rivers in an instant, shining bright white in the reflected sunlight. I have never seen anything like that. As I looked down on that place, I had this thought: The people there are living in the jungles of the South Pacific, probably fishing for sustenance in those rivers, and there are probably a lot of people there who have never had any modern technology. What would they think if they knew an astronaut was floating overhead in outer space, looking down and taking a picture of them? It was a poignant moment for me.

The color that stands out most on Earth is the blue of its oceans, but white was a close second. I was in space for several months of the Northern Hemisphere winter, and during that time Canada and Russia were white, their snow and ice going on and on for thousands of miles. In Canada there are vast, flat plains covered in snow, in the west the Rockies wrinkle the land, and as you move east the great forests make patterns of rivers and valleys, including the ancient Manicouagan impact crater in Quebec that is one of the most recognizable features on our planet from space. The Hudson Bay is huge, and it has a pattern of ice that to me often looked like clouds—there were many times that I couldn't tell if I was looking at ice floes, clouds, or just snowy ground, all the way over to the Atlantic. Siberia, on the other hand, was even more vast than Canada. Russia's time zones famously cover half the Earth, and when seen from space, it spreads out from one horizon to the other, nothing but white snow and ice and clouds. In the southern part of Siberia is the largest freshwater lake on Earth, Lake Baikal. Visiting this crescent-moon-shaped lake is definitely on my bucket list. Flying a little farther to the east, you come to the strangest place on Earth, in my opinion: Kamchatka. It reminded me of Venus. It was always covered in snow and is full of volcanoes. There always seemed to be at least one erupting, a black stream of smoke and ash clearly visible from the ISS. The islands near there, especially Sakhalin, are also very strangely shaped, with otherworldly sharp curved patterns that looked more like an ancient samurai sword than a body of land.

Because the ISS orbits between 51.6 degrees north and south latitude, the extreme polar regions were only visible in the distance. I saw the Antarctic landmass a few times from space, small white peaks jutting up out of the terrible Southern Ocean weather, but the white of Alaska and the Aleutian Islands was much more visible. They were a string of volcanoes jutting up out of the North Pacific that went on and on for hundreds of miles, usually buffeted by those bizarre storms in the region.

During the day, you can navigate by landmass patterns or colors. But at night it's by light, the most obvious being city lights. One of the first night passes I saw was over the Middle East and I noticed two things very quickly. First, that place was small. Israel was right next to Jordan, Egypt, Syria, and Lebanon, all in a very small piece of terrain, and I wondered, "Why all the fuss? Why is there so much conflict in that very small area? You'd think we could get along better." I'm a realist and I know there are reasons for conflict, but when you see Earth from space it doesn't seem like there should be.

The next thing I noticed was that most city lights are yellow, but some are pure white and others blue, with occasional red dots. There are different technologies that make those lights—mercury vapor, sodium, halogen, fluorescent, LED—and each of those emits a different color. In the Middle East, you could tell the country by the color of light. In some cities, however, you see different colors in different neighborhoods. It was so fascinating to see these details of daily life from my orbital perspective.

Most surprising were fishing-boat lights at night. I could tell where I was by the color and pattern of fishing boats in Asia. There were several groups of boats that always seemed to be off the coast of Korea, making a swirl pattern, almost like a comma. There must have been hundreds of those squid boats, all with very powerful spotlights. Japan also had some consistent patterns of fishing boats off its coast. They moved and appeared and disappeared with the season, so you could tell not only where you were, but also when it was, based on the fishing-boat lights. Most fascinating of all were the green boats in the Gulf of Thailand. It was unique, a massive fleet of green-lighted boats there in the sea, plainly visible from space, night after night. It quickly became obvious where I was if I saw a green field of lights at night—southeast Asia, next to Thailand. Another bucket list item—going fishing at night in Asia to see this vast searchlight armada firsthand.

Colors during the day and city lights at night were all things that I never imagined would help me learn my way around Earth. But they all gave me a newfound appreciation and knowledge of our one and only home in the universe, our beautiful planet, Earth.

BAD BOSSES

Silly Rules and Bureaucratic SNAFUs

One of the low points of my mission was having my butt chewed out by the deputy chief of the astronaut office, via email, for not paying attention to the exercise constraints on our daily schedule. These were times when we weren't allowed to run on the treadmill because the robotic arm was scheduled to be in motion on the outside of the station. We assumed the ground controllers didn't want vibrations from exercise to interfere with the arm's movement. Getting yelled at for this reason was a bummer because I had actually made this a big emphasis item for my crew; it was something that we were aware of and tried to avoid.

So, I gathered our crew together and reminded everyone once again: "Folks, remember to check the electronic schedule before exercising, and be sure there are no constraints, or Houston is about to get really mad and take away our freedom to exercise when we want to." We brainstormed about ways to avoid such situations—call Houston before beginning every exercise situation (a real pain), putting a big tag or something on the machine that you had to remove before exercising that would remind you to check, etc.

Then one of my crewmates asked a simple question: "Why do we even have this restriction to begin with?" We thought it was silly and an overabundance of caution, but that was par for the course. After a few minutes of discussion, I grabbed the microphone and asked a very innocent question: "Houston, we have a quick question about exercise constraints. Can you please confirm the rationale behind the 'no running on the treadmill during robotic ops' constraint? Is there a concern that vibration from the treadmill may damage the robotic arm?"

Awkward silence. Pregnant pause. I imagined that down in mission control, folks were doing rock, paper, scissors to see who would have to call us with the answer.

The reply came up sheepishly. "Uh, station, no, actually the concern is that the motion of the arm would cause a problem for the treadmill." I knew NASA was sometimes ridiculously overcautious in its safety rules, but come on!

They were actually worried that the robotic arm, moving at a blistering pace of 1 cm/second, would cause so many vibrations as to damage the treadmill.

It turns out that the space station program had not been willing to spend the money required to do the engineering analysis to see if arm motion would damage the treadmill. Well, I set up a video camera to film the treadmill one day while the arm was moving to demonstrate that the treadmill wouldn't come flying off its hinges when that small arm moved. Of course there was absolutely no discernible motion from the fragile [sic] treadmill. No $$$ or engineering analysis required. I don't think the bosses much appreciated my video, but I thought it was funny.

This may have been the most absurd thing I heard during my tenure at NASA. They were actually worried that the robotic arm, moving at a blistering pace of 1 cm/second, would cause so many vibrations as to damage the treadmill. This was a device designed for 200-pound astronauts to run on, which of course would really bang it around, and, in any case, it was protected by a vibration isolation system, floating on springs. To spend more than ten seconds talking about such a threat defied common sense and logic, yet NASA had given the crew a constraint that required mission control to waste time tracking it on our schedule and also took the time of astronaut managers to write sternly worded emails to protect us from a problem that posed a 0.00 percent threat to anything or anybody.

We got a good laugh out of the experience—but also shook our heads at a culture and decision-making process that would think it was OK to constrain the crew from an activity that was obviously not even a remote threat. And then chew them out for not following said constraint. We saluted smartly and moved on with our lives, making sure not to exercise when our schedule was flagged. But the whole episode brought up some important lessons for me.

First, sometimes people are unreasonable, and there are rules you don't agree with. Often you just have to deal with them and go about your life, because it's not worth the trouble or stress to think about. Second, if something really needs to be changed, try to fix it through proper channels. Don't just complain—give the boss a solution. During my postflight debrief I made this exercise constraint one of my top issues, and management agreed that it was silly and needed to be resolved. However, the last time I checked, two years after the fact, the restriction was still in place. Finally, if you're in a position of authority and there are rules that make absolutely no sense, change them. Don't add to the expense and frustration of your people unnecessarily; just make a command decision and do the right thing. If you have the power to get rid of waste, get rid of it. Everyone will respect you for it.

IN SPACE NO ONE CAN HEAR YOU SCREAM

An Ammonia Leak Threatens the Station's— and the Crew's—Existence

For all the emergency training I went through as an astronaut, I never expected to be holed up in the Russian segment of the ISS, the hatch to the US segment sealed, with my crew waiting and wondering—would the space station be destroyed? Was this the end? As we floated there and pondered our predicament, I felt a bit like the guy in the Alanis Morissette song "Ironic," who was going down in an airplane crash, thinking to himself, "Now isn't this ironic?" This is how we ended up in that situation.

Every space station crew trains for all types of emergencies—computer failures, electrical shorts, equipment malfunctions, and more serious fire and air leak scenarios. However, on the International Space Station, the most dangerous of all is an ammonia leak. In fact, our NASA trainers used to tell us, "If you smell ammonia, don't worry about running the procedure, because you're going to die anyway." That sure instilled confidence.

A few months after arriving in space, we were having a typical day. My crewmate Samantha Cristoforetti and I were each in our own crew quarters, going through email and catching up with administrative work, when the alarm went off. The sound of the ISS alarm is exactly what you would think a proper space alarm should sound like—a cross between a *Star Trek* alarm and a sci-fi B-movie klaxon. When it goes off, there is no doubt that something significant is happening. Sam and I both popped our heads out of our respective quarters and glanced at the alarm panel.

When I saw the ATM alarm lit up, my first thought was, "Atmosphere—there must be an atmosphere leak." The ISS had occasionally had an air leak false alarm over its fifteen-year history, and I thought it must be one of those. However, that is not what ATM means—it stands for toxic atmosphere, most probably from an ammonia leak. Significantly, this alarm was going off for the first time in ISS history. My brain couldn't believe it, so I said to Samantha, "This is an air leak, right?" To which she immediately responded "NO—*ammonia leak*!"

Jolted back to reality, we jumped into action. Gas masks on. Account for everyone; we didn't want anyone left behind. Float down to the Russian segment ASAP and close the hatch between the US and Russian segments. The US segment uses ammonia as a coolant, but the Russian segment doesn't, so the air should be safe there. Remove all clothes in case they're contaminated. Nobody smelled ammonia, so we skipped this step! Close a second hatch to keep any residual ammonia vapors on the American segment. Get out the ammonia "sniffer" device to make sure there isn't any of that deadly chemical in the atmosphere on the Russian segment. All clear. Then, await word from Houston. . . .

Fifteen long, suspense-filled minutes later, we got the news—it was a false alarm. We let out a collective sigh of relief; the station wouldn't be dying today! Whew. Similar to frequent fire alarms and rare air leaks, ammonia leak was just added to the collection of ISS false alarms. We put away the ammonia detector, floated back to the US segment, and started to clean up the mess that we had left floating in midair when that alarm went off.

Then we received an urgent call. "Station, Houston, execute ammonia leak emergency response, I say again, execute emergency response, ammonia leak, this is not a drill!" Pretty unambiguous. Only this time the warning had come via a radio call, not via electronic alarm. After the false alarm I knew that an army of NASA engineers were in mission control, poring over every piece of data they had, trying to determine if this had been a false alarm or the real thing. Now that mission control had confirmed that it was an actual leak, there was no doubt in my mind that this thing was real. No way all those

NASA engineers got this call wrong. Having worked in mission control for nearly a decade myself, I had complete confidence in our flight director and flight control team. When they said, "Execute ammonia response," I put the mask on, shut the hatch, and asked questions later.

It was like a scene out of *European Vacation*—"Look kids! Big Ben!"—or maybe *Groundhog Day*. Oxygen masks activated—check. US segment evacuated with nobody left behind—check. Hatch between US and Russian segments closed and sealed—check. Get naked—nope. No ammonia in the Russian atmosphere—check.

By this point, we had run the ISS ammonia leak procedures twice within an hour of each other. We had a quick debrief as a crew to discuss how we handled the emergency, what checklist steps were missed, what could have been done better, and what we needed to report to Houston. By this point, it was very obvious that there would be a lot of meetings happening in Houston and Moscow and that everybody in the NASA chain of command would be aware of our predicament.

Very quickly the gravity (pun intended) of the situation hit us. Using ammonia as the coolant for the American half of the ISS had worked well for decades, but we were acutely aware of its danger. Thankfully, the engineers who designed the station did a great job making a leak extremely unlikely, but the possibility was always there. On the other hand, the Russian glycol-based coolant is not dangerous, which is why the whole station crew would safe-haven there in the event of an ammonia leak.

Besides the danger of the crew breathing in toxic fumes, there was a risk to equipment. The ISS has two ammonia loops, a series of tanks and pipes that carry heat from the station's internal water loops to the external radiators. If one leaked out to space, there would still be a second available to cool equipment. It would be a serious loss of redundancy for the station, especially given that there is no longer a space shuttle to restock the station with the massive ammonia tanks needed to fill a loop. It would be ugly, but survivable.

What is not survivable, however, is having that ammonia leak to the inside of the American segment. First of all, if the entire contents of an ammonia loop came inside the station, it would probably overpressurize and pop the aluminum structure of one or more of the modules, like a balloon being overfilled with air. Mission control could avert this problem by venting the ammonia to space—we would lose the cooling loop, but it would prevent the station from popping. Months after returning to Earth, I learned that Houston had been seriously considering that option during our emergency, and it was only averted because of a tough—and ultimately correct—call by our flight director. That's why those guys get paid the big bucks—they are some of the smartest and most competent people I have ever worked with. However, even if you averted a catastrophic "popping" of structure, there would still be the problem of ammonia in the US segment.

If even a small amount of ammonia were present in the atmosphere, it would be difficult, if not impossible, to remove. The only scrubber we had was our ammonia masks, so theoretically you could have an astronaut sit in a contaminated module, breathing the contaminant out of the air and into the mask filter, and over time enough of this scrubbing would lower the ammonia concentration, but as the poor astronaut sat there cleaning the air he would also be covered in ammonia, and convincing his fellow crewmates on the Russian segment to allow him back to their clean air would be problematic, to say the least. There would need to be some sort of shower and cleaning system to completely clean him up, which of course doesn't exist in space. It would be a similar situation to soldiers in a chemical warfare environment, or the Soviet soldiers in the recent miniseries *Chernobyl*. Dealing with a toxic environment on Earth is difficult enough, but in space it would be nearly impossible. The reality is that an actual leak into the American segment would make a significant portion of the ISS uninhabitable, and if there were no crew there when the equipment broke down, there would be nobody to fix it.

A real ammonia leak would eventually lead to the slow death of the US half of the ISS, which would then lead to the end of the entire station. We knew this and spent our afternoon staring at each other, wondering out loud how long it would be before they sent us home, leaving the space station uninhabited and awaiting an untimely death.

In the midst of that unexpected and extreme situation, something extraordinary happened. If you have seen the news over the past decade, you may have noticed that there has been a bit of tension between Russia and the West: Ukraine civil war. Sanctions. Crimea. The shooting-down of a Dutch airliner in 2014. There was plenty of tension to go around. In that context, we received an unexpected radio call from one of the top officials in the Russian government: Mr. Dmitry Rogozin, who had actually walked me out to my Soyuz rocket on launch night. He was one of Mr. Putin's top deputies and was right in the middle of the tension between Russia and the West. He told us that they would work with us. "Our American colleagues can stay as long as they need on our segment," he said. "We will work through this emergency situation together."

> It was like a scene out of a space horror film; all the things that we had let go in the mad rush to escape the leak were floating about randomly, as if a ghost were carrying them.

This was an example of how people can and should work together—in space, on Earth, in a family, in a business relationship, or between nations. We were concerned about working through this potentially deadly problem without the ridiculous encumbrance of grandstanding politics. I considered my Russian crewmates to be my brothers and sisters, and I immensely enjoyed flying with them. The time I spent hanging out with them in the Russian service module, eating dinner after the work day, was a highlight of my time in space. And here we were, faced with a deadly situation in that same service module, working together to survive and save the station. I'm no idealist, and having spent thirty-plus years in the Air Force, I know that sometimes strength is required in international relations, but the way we handled that ammonia leak is the way things should be done down here on Earth.

Later that evening, we received a call from Houston. "Just kidding, it was a false alarm." That was a *huge* false alarm. It turned out that some cosmic radiation had hit a computer, causing it to kick out bad data regarding the cooling system, and it took Houston hours to sort out what was really happening. Because that call from Houston had told us that it was a real leak, we all believed it—we knew that the folks in mission control were some of the best engineers in the world and that they would be 100 percent sure before making a call like that. So we were *very* relieved to get that call.

"False alarm" notwithstanding, they asked us to wear gas masks when we reopened the hatch and returned to the US segment to sample the atmosphere, "just in case." Which made me chuckle. So a crewmate and I put our gas masks back on, ammonia detection equipment in hand, and floated down into an abandoned spaceship. It was like a scene out of a space horror film; all the things that we had let go in the mad rush to escape the leak were floating about randomly, as if a ghost were carrying them. They looked as if they had been drifting aimlessly for centuries, abandoned by a long-dead crew, even though it had only been a few hours. Luckily, the atmosphere was clear and our masks came off shortly after returning, and life on the station was back to business-as-usual within a day or so.

For me, the big lesson from this whole drama wasn't how to improve spaceship cooling loop design or procedures. Or "Should we or shouldn't we take our clothes off once on the Russian segment?" (My advice—don't.) It was the example of how people should work together to solve important problems, leaving petty political bickering behind. That is exactly what we did and what the space program in general has done for many decades. The vacuum of space is a harsh and unforgiving environment, and it doesn't care what country you are from or what your ideology is. Unless you approach spaceflight focused only on getting the job done and working as a team, you risk dying.

And that, my friends, is a lesson that we would do well to learn down here on our home planet.

IT WAS A LONG 200 DAYS

Do ISS Astronauts Make Whoopee?
(What Everyone Wants to Ask)

The one question that everybody wants to know the answer to, but many are afraid to ask, is "Have you, uh, you know? Or has anybody? You know. . . . In space?"

To answer this, I'll give the same answer I gave Neil deGrasse Tyson when I was a guest on his TV show at the Hayden Planetarium in New York City. The format of his show is to have a segment filmed in front of a live audience, then Neil talks informally with the audience off camera during intermission, and finally he comes back to finish the show. During the intermission segment, Neil turned to me and said, "Terry, come on, be honest, did you? Or has anybody?"

And my answer was . . .

Before I get to the answer, let's just talk through logistics. The Soyuz is like you and your two best friends in the front seat of a Volkswagen Beetle. It just ain't gonna happen in that small space. There is literally NO privacy whatsoever. Also, you are in the Soyuz immediately after launch, and typically fly a four-orbit rendezvous, which means that it only takes six hours to get to the station. Much of that time is busy with procedures, and all of it is spent in your Sokol spacesuit, a big bulky pressure suit that takes ten minutes to get out of. Which renders the practicality of getting busy impossible. Even more important, nobody is in the mood during the few hours after leaving Earth. Your brain's vestibular system is still shaking its head and asking itself, "What the heck just happened to me and where is gravity???!!!"

There is the possibility of a two-day rendezvous after launch, in which case you'd have some spare time to do the deed. Crews actually get out of their Sokol suits and slip into something more comfortable, like a polo shirt and

Velcro-laden Bermuda shorts. But—and this is a big but—your brain's vestibular system is still completely confused, which makes getting in the proper mood for said activities exceedingly difficult, if not impossible. What's more, the Soyuz would be flying in a rotation mode, spinning like a chicken being roasted on a rotisserie spit, to prevent the sun's energy from overheating one side. While in this rotation mode, your brain would feel even worse from the spinning. What's more, if the crew is bouncing around inside and banging against the walls, the very tiny Soyuz would be knocked out of attitude, triggering an alarm, followed by a stern warning from mission control Moscow, in the Russian language no less. None of those things are conducive to . . .

The space station, on the other hand, is big. Very big. Larger than a 747 big. Plus there are only six or fewer of you there for half of the year, so one would think that there would be time for extracurricular activities.

The space shuttle, on the other hand, was a bit larger. It was more like you and six of your best friends in the kitchen, or maybe the kitchen and laundry room, for two weeks. So . . . probably ain't gonna happen. Shuttle flights had all of the same medical concerns as the Soyuz, though its missions were frequently longer than a week, which means the crew would get over their initial dizziness. Here's the bottom line—the shuttle did fly 135 missions, and where there's a will there's a way, but I seriously doubt that anything interesting ever happened.

The space station, on the other hand, is big. Very big. Larger than a 747 big. Plus there are only six or fewer of you there for half of the year, so one would think that there would be time for extracurricular activities. And the aforementioned nausea and vestibular problems are gone after a few days. Everyone feels very normal and well adjusted for the vast majority of those six months.

The physics of this problem would be easily solvable. There are handrails in every corner of the ISS that are used by crews to move around using their hands and also to hold themselves in place with their feet. So, it's pretty easy to stabilize yourself. And in weightlessness lots of things are possible. And people tend to be very creative, out of either desire or necessity. In fact, the

DEALING WITH
A DEAD CREWMEMBER

If a Fellow Astronaut Expires

NASA is always planning for the unlikely in order to avert disaster. Virtually every contingency is thought of, trained, backed up, and mitigated. But at the end of the day, crews are human, and not only do people sometimes make mistakes, they don't last forever. One change in crew demographics brings attention to this reality. The early astronauts in the 1960s were rarely over forty years old, but today we routinely fly people in their fifties and even occasionally in their sixties.

Which leads to the possibility that someone will die while in space, either in Earth orbit or during a mission to the Moon or Mars. Therefore, it is a perfectly logical question to ask what would happen to the body in such a situation. I don't recall being specifically trained on this contingency—the coldhearted word NASA uses for an "off-nominal situation," which is the NASA term for "something wrong." But as astronauts we realized that this was a possibility, and though it wasn't often discussed, I'm sure that my colleagues would handle this situation professionally and compassionately.

That being said, there are a lot of practical details to consider. First and foremost: What caused the death? If there were a fatal accident, or if one person who had a known medical condition died, then the medical situation would be well understood, and life on the station would go on after dealing with the body. However, if there were a serious problem on the spacecraft that killed the individual, such as an ammonia gas leak or toxic fumes, the first priority would be solving the malfunction to prevent more astronauts from suffering the same fate.

Even worse than someone having an untimely heart attack or breathing in toxic fumes would be if your fellow astronauts started dying from an unknown cause. That would definitely grab everyone's attention and probably give the surviving crew a powerful motivation to return to Earth as soon as possible. Which, depending on the cause of the mystery deaths, might not be the best course of action. Think *Alien*, *Invasion of the Body Snatchers*, or any one of a thousand "infection from outer space destroys humanity" movies. It's conceivable that an unknown plague affecting a crew in space would be dealt with by simply leaving the crew in space. They would have the rest of their lives to solve the mystery, if they were to have any hope of ever being allowed to return to Earth.

Regardless of the cause of death, there are some practical concerns that would need to be addressed. There are basically two ways to deal with a human body in space. One option is to give a burial at sea, as it were. The other option would be to return the body to Earth. Both of these options would be discussed among the crew in space, the family on the ground, and management on Earth. This decision would certainly be the hardest one any of those involved would ever make in their lives.

The burial-at-sea option would require taking the body out of the airlock on one final spacewalk and floating it off into space. A practical concern would be how to get the body outside. You would have one or two crewmembers in the airlock tethered to the deceased astronaut. They would take the body out of the hatch, release the tethers, and slowly push their comrade away in an aft and radial down direction—toward the back of the station and down toward the Earth. This would ensure that gravity and orbital mechanics would pull the body away from and below the station, eventually to burn up in the atmosphere. A lightweight, large-surface-area mechanism attached to the body would act as a sail, increasing the drag from Earth's tenuous atmosphere and speeding up its orbital decay and inevitable re-entry.

If the calamity happened during transit to or from another planet, you would have to pick a direction that would keep the body from recontacting the spacecraft, and preferably keep it from interfering with the re-entry that the rest of the crew would perform when they got to their destination planet.

If it occurred on the surface of the Moon or Mars and the body were to be left in place, you would have to find a suitable location outside for burial—the first extraterrestrial cemetery. I often show a photo of the Earth during my own talks and remark, "Every human in history was born and died down there on Earth." When the first human dies on another planet, this anecdote will no longer be true.

Another question—how do you dress the deceased? You would certainly want the body contained, probably by at least two methods, to avoid a further mess. Initially, it would go into one of the body bags we have stored on the ISS for such a contingency. It could then be placed in a standard spacesuit, but this option would be highly undesirable. First, EMUs are expensive, there are only a few operational suits on board the station, and it would be really difficult to move an uncooperative person around in that giant suit. A better option would be to keep the individual in the body bag and then put that into a large storage bag, called MO-1 or MO-2 bags, about the size of a dishwasher. Using the deceased's launch and entry suit from the Soyuz/Boeing/SpaceX capsule is another possibility. The problem is that those suits inflate in low pressure and so would be really difficult to get out of the airlock hatch. Think kid in snowsuit in the movie *A Christmas Story*.

A final burial-at-sea option would be to send the body outside alone, through an equipment airlock, without a crewmate in attendance. The body would be released robotically, in the same manner that space station astronauts launch small satellites. It would slowly fall away into the silent abyss, fading into the distance as it floated into the black sea of the universe, humanity's first burial in outer space.

The other significant option is to bring the body back to Earth. This might be an appropriate course of action if the cause of death were undetermined but seemingly benign, or if the family so desired. In this case, the body would be placed into its launch and entry suit and then strapped into the capsule seat for the deorbit burn and landing. On the Soyuz each of the three crewmembers has their own custom-molded seat, so if the deceased had been the vehicle commander or flight engineer, their seat would have to be relocated to the second flight engineer position, where there aren't many tasks to do during landing. This could prove impossible if rigor mortis set in and made it impossible to fit the body in the seat. A final intriguing option would be to put the body in a cargo vehicle. Most of the cargo ships that depart the ISS burn up in the atmosphere, full of trash, but the SpaceX Dragon actually splashes down in the ocean under parachute, and that would potentially be a better option to safely return a body to Earth rather than flying the deceased in his or her normal seat in the capsule.

This is a grim topic, and thankfully no space traveler has yet died while in space. But if humans continue to leave Earth in the coming decades and centuries, it will eventually happen.

ROBOTIC CREWMATES

Remote Work Outside the ISS

Robotics has become a miracle field in the twenty-first century. Most manufacturing is accomplished by robots, drones are everywhere, military and emergency responders have robotic helpers for dangerous situations, and even some surgery is performed by robots. Progress in this field will continue to accelerate exponentially, and people at the end of the twenty-first century will look back at us and chuckle at how primitive our technology is today. But when I flew into space at the beginning of the century, flying a space robotic arm was still pretty cool.

Actually, there are multiple robotic arms in space. The shuttle had the original Canadarm, a six-joint manipulator built by, you guessed it, Canada. It was used since the early days of the shuttle program to pull satellites and equipment out of the payload bay (the shuttle's cargo hold) and deploy them into space. It was also used to grapple free-flyers, orbital satellites that the shuttle would approach, grapple with its robotic arm, and either repair or stow in the payload bay for return to Earth. This was how the *Hubble Space Telescope* was repaired—grappled by the robotic arm, attached to the aft end of the shuttle, where spacewalking astronauts repaired and replaced its broken instruments, and then pulled out of the payload bay by the arm and released back into space, where it remains today, orbiting the Earth.

Later in the shuttle program, the arm was used to pull massive ISS modules out of the payload bay and hand them off to the station's much larger arm, Canadarm2. Guess where that was made? This was one of my main tasks during STS-130, the only shuttle flight to bring up two station modules at once—Node 3 ("Tranquility," the large living module) and the Cupola (the small, seven-windowed observation module). The Cupola was attached to Node 3 for launch, so my task was to pull that combination out of the shuttle and hand it off to my colleagues on the ISS who were waiting to grapple it with the big arm, after which they attached the whole assembly to an empty docking location on the Node 1 module.

This was by far the most stressful moment of my entire shuttle mission. Node 3 and Cupola were launched while firmly attached to the payload bay by two keel pins, or clamps, on the bottom of Node 3. In order to remove the modules from the shuttle, those clamps had to be released. However, a lot of tension had built up because of the stress of launch, so when they were released the entire 35,000-pound assembly lurched and floated toward the right side of the shuttle, coming to rest a few inches from the wall of the shuttle. My task was to manually lift these modules straight up and out into free space, like a crane lifting construction equipment. But after the module had moved so close to the payload bay, there was no room for error without bumping our precious and massive cargo into the shuttle's thin walls, possibly damaging our ride home. At least all of my friends and family and colleagues and people all over the world would be watching the majestic moment when our massive cargo was gently floated into space and handed to the station's waiting arm in a slow-motion orbital ballet. By me. All I had to do was thread the needle of this extremely narrow corridor with barely a few inches of tolerance from *Endeavour*'s starboard wall. What's the worst that could happen?

Long story short—I was able to do it, with a lot of fellow astronauts back on Earth probably saying to themselves, "Man, I'm glad that's not me." What a huge sigh of relief I gave when the bottom of Node 3 had finally cleared that very fragile-looking shuttle wall and been successfully grappled by the station's arm. For me and my fellow robotics operator Kay Hire, that wasn't the end of our job—we still had to float over to the station and do several more robotic operations with the big ISS arm—but this was a significant task completed and a huge morale booster.

Best of all, I had avoided getting a new nickname from my fighter-pilot buddies. Some possibilities that were floating in my head: "Bump," "Crash," "Robo," "Ding," etc. Air Force fighter pilots are fond of giving call signs that emphasize a fault or defect, as opposed to gratuitous call signs that our Navy colleagues are fond of giving, like "Maverick," "Iceman," and "Cougar."

Another important task that the shuttle arm fulfilled during the final twenty-two shuttle flights was performing an inspection of the orbiter's heat shield on the bottom of the vehicle. This was a critical lesson learned after the *Columbia* accident on STS-107, when a fragile heat shield was damaged during launch and neither the crew nor ground had a good idea of the extent of the damage. NASA made the fateful decision to ignore that damage and not even try to take a photo of any potential damage, and the crew was killed as *Columbia* burned up during re-entry because of the hole in their heat protection. After that terrible tragedy, NASA developed the OBSS (NASA acronym for inspection boom). It was a 50-foot-long, massive extension of the shuttle's arm that had several cameras and a laser sensor. On the second day in space, before docking with the ISS, we grabbed the OBSS with the arm, and with cameras running we slowly moved it along the leading edge of the orbiter's wings and along much of the tile on the bottom of the vehicle to verify that no damage had occurred during launch. This whole procedure was repeated again on the day before landing to make sure that no meteorites or orbital debris had damaged the heat shield during our two weeks in orbit. The process was very tedious and time-consuming, because the sensors needed to move slowly so they could get very detailed images of the heat shield. Most importantly, we needed to be very careful not to bump into a fragile part of the orbiter, potentially damaging critical thermal protection.

> Anybody who has used an Xbox or PlayStation controller could figure it out pretty quickly. Alas, most senior (i.e., old) astronauts are not familiar with such things.

During STS-130 I was tasked to do this inspection, together with my crewmate "Stevie-Ray" Robinson. It was scheduled to occur as we were flying over South America at dusk, and I still vividly remember those spectacular thunderstorms. It was dark enough that we could see the brightness and detail of lightning flashes, but there was still enough sunlight to see the fading colors of the Amazon, along with the white and gray shading of the clouds. It was a sight that I never saw again, not even during my 200-day-long mission, which taught me an important lesson—enjoy every amazing moment in life, because you don't know if or when it will happen again. While I was gawking at our beautiful planet, poor Steve was having to fly the arm by himself. I still owe him for that.

The space shuttle's arm was what is known as a "six-degree-of-freedom arm" (or 6 DOF). Much like a human arm, there was a shoulder joint that could yaw and pitch, an elbow that was pitch only, and a wrist that could yaw, pitch, and roll. Which basically meant it could move like your arm on your body. On the other hand, the station arm was much bigger and also had an extra degree of freedom—its shoulder joint was exactly like its wrist joint, so they could each roll, pitch, and yaw. This "7-DOF" configuration meant there were a lot of possibilities for arm motion, but it also meant that you could get really confused. So, we would usually lock one of the shoulder joints while the arm was moving to turn it into a 6-DOF arm like the one on the shuttle. The problem with letting all seven joints rotate while moving the arm was that it would swing unpredictably, especially the elbow area, and we really wanted to prevent unexpected arm motion so that it didn't accidentally bump into structure.

The critical reason for giving the arm seven degrees of freedom was that it could inchworm across the station: The wrist would grab on to a grapple fixture and become the new shoulder, and the old shoulder would let go and move out into space, becoming the new wrist. We needed this capability because the station was so large; no single arm could have provided coverage to the whole exterior without being able to move itself around.

There is also a robotic arm on the Japanese module that is used to pull equipment from their airlock, which is located inside the station, and stow it on an external platform, exposed to the extreme environment of open space. This is a very useful capability that allows a lot of science to happen without the overhead and danger of astronaut spacewalks. The Russian segment also has a robotic arm that is manually controlled and especially useful during their spacewalks. Finally, there is Dextre, also known as SPDM, the Canadian acronym for "fine manipulator." Although it is quite massive, it has several smaller arms and hands that can perform certain tasks that the big arms cannot, such as pulling small equipment out of the trunks of cargo ships, or manipulating bolts or small equipment.

There were two basic jobs for all of these arms. First, to move equipment from point A to point B. This was the main task during the years of station assembly; we were constantly moving massive modules around and assembling them. The second task was to reach out and grab approaching cargo ships—they would fly in formation, hovering about 30 feet below, while the crew would fly the arm down and grapple the vehicle and then attach it to a free docking port. Most arm motion today is controlled by the ground in order to free up precious crew time. But grabbing those free-floating vehicles is something that the crew still does because it is such a time-critical operation, and it is one of the most fun, dynamic tasks that ISS astronauts have, breaking up the monotony of repairing equipment and performing science experiments. Well, we also have launch, landing, rendezvous, and spacewalks, so I hear you, monotony for astronauts really isn't that bad.

There are basically three techniques to fly a robotic arm in space: manual, automatic, or joint control. Manual was my favorite but was typically only used for grappling visiting vehicles. Anybody who has used an Xbox or PlayStation controller could figure it out pretty quickly. Alas, most senior (i.e., old) astronauts are not familiar with such things. In order to move the arm, you take the THC (controller in your left hand) and push forward, and the arm moves forward toward the target. You could move it left/right/up/down and get corresponding motion from the arm's end effector—this motion is called translation. You also use the RHC (controller in your right hand) to pitch, roll, and yaw the arm to be aligned with the grapple fixture on the cargo ship—this motion is called rotation. Once the hand of the arm moves over the target's grapple fixture, you squeeze the trigger, some steel cables inside the hand pull tight, and the arm attaches firmly to the target—the cargo ship floating below the station, hardware to be moved, or the new home for the arm, as its hand becomes its new shoulder and vice versa.

Much of our arm motion was done in automatic mode. In this mode, flight controllers would calculate a position and orientation for the arm to go to and come up with computer commands to move it there. Then either the crew or ground could command the arm to move to the desired position and attitude. It wasn't very exciting, but you still needed to have good camera views to make sure that all parts of the arm (wrist, elbow, booms, shoulder, etc.) were clear of structure. The big concern in any arm motion was accidentally bumping into something, and this required constant vigilance by looking out the window in the Cupola, monitoring the many external video cameras available, and following the expected trajectory on a laptop.

The third method of moving the arm was by directly controlling each individual joint angle. Imagine moving your own arm by pitching your shoulder up 20 degrees, yawing it right 20 degrees, bending your elbow down 50 degrees, yawing your wrist 10 degrees, and rolling it 30 degrees, etc. You get the point. You can point the arm exactly how you want it by telling each joint what angle to go to. This is tedious and time-consuming, but precise. I always enjoyed challenging myself in the simulator to control the arm with joint angles only, trying to grapple a target or rescue a stranded spacewalking crew-member with the arm, trying to move him quickly to the airlock before his oxygen supply ran out. These far-fetched and difficult scenarios in the simulator were unlikely to ever happen in real life, but they were a critical part of my training because they forced my brain to fully understand how the arm worked and to visualize complex 3D problems. Those tough exercises made me a better arm operator than any other training I did.

Humans and robots are going to continue to work together more and more in the future, both here on Earth and in space. Flying these twentieth-century-vintage robotic arms was a blast. I hope that robots in the future help crews in space as well as us poor earthlings—and don't fulfill the dystopian Hollywood visions of HAL or the Terminator.

PHONES, EMAIL, AND OTHER HORRORS

Communicating with Earth (Slower than Dial-Up)

How do astronauts keep in touch with friends and family back on Earth? The answer is both simple and complicated. There are lots of ways to connect with loved ones (and not-so-loved ones), which forced me to ask an introspective question. Did I really want to connect? In many ways I looked at that half year off the planet as a once-in-a-lifetime opportunity to disconnect. No internet. No texting. Limited phone and email. For me, it was the imperative to disconnect that was a blessing.

There were several ways to communicate with our fellow earthlings stuck on our planet. The most common method was by email. Good old Microsoft Outlook. For the first fifteen-plus years of the ISS, email was synchronized thrice daily. So you'd write an email, wait a few hours for it to be sent, then wait a few more hours until the next sync period happened, and hope there was a reply. Or not, depending on the subject of the email! If there were no reply, you'd have to wait hours for the next sync. This process went on for years, rendering email a not-very-efficient method of communication. Luckily, a few years ago the station program transitioned to a new system of continuous email syncing, where as long as the station is in satellite coverage, email accounts constantly synchronize. This allows emails to be used as a poor man's text. You still have to float back to your crew quarters' laptop to check email, and you have to have appropriate satellite coverage (roughly 90 percent of the time), and your buddy on Earth has to be checking his email religiously. But if all of those conditions are met, voilà—space iMessage!

There is also the possibility of logging on to the internet while on board the ISS. It is very slow, with speeds reminiscent of the good old dial-up days.

It is only available when the appropriate comm satellites are in view of the ISS. And it requires a relatively painful log-in process. Despite all of these limitations, some astronauts love using the internet because it allows them to log on to their social media accounts so they can tweet directly, without having to email photos and quotes to a middleman on Earth who would post them to the appropriate social media account. It also allows you to surf the web, looking up such obscure questions as "How tall are the sand dunes in Namibia?" and "Where is the south magnetic pole located?"—two questions I asked the Google while in space. However, I quickly concluded that whatever strange facts I could learn through the internet weren't nearly worth the pain required to log on and then wait forever for a Google query to get processed. I looked at my flight as a blessed opportunity to mostly go without the internet for 200 days, and only rarely logged on when absolutely necessary. It was exactly the detox that most of us could use.

There were several other methods of communicating besides email. The most common was a telephone system that used voice over IP technology, appropriately called IP Phone. Today, there are many apps that do this exact thing on your smartphone—WhatsApp, Viber, FaceTime, Signal, Vonage, and the list goes on. But back in the 2000s, when we first got this system, we thought it was hot stuff. Once again, it depends on those comm satellites, which warrants a brief description. The ISS doesn't normally communicate by sending radio signals directly to ground stations on Earth; it sends them to TDRSS (pronounced *tea druss*) satellites operating in geosynchronous orbit, which means they are usually 30,000 miles or more away from the ISS. There are two radio frequencies that we use with TDRSS: S-band and Ku-band. S-band is for voice communication and very low rate data. Ku is a much higher data rate, and we use it for things like payload data, video, and the IP Phone system. Our phone software was on our laptops and unfortunately didn't have an address book system, so I had to memorize everyone's phone number, just like back in the twentieth century.

I like to joke that this is the most perfect phone in the world—you can call anywhere for free, and nobody can call you. Another advantage is that it suddenly and unexpectedly hangs up. As the ISS flies along at 17,500 mph, it jumps from one TDRSS satellite coverage zone to the next, and when it flies out of view of the satellite it is using, there is Ku-dropout and the phone suddenly hangs up. No long, drawn-out goodbyes, just "click" and the call is over. Plus, the maximum length for which you might have any one particular TDRSS satellite in view is about forty minutes, so every phone call is naturally limited. Like I said, it's the best phone in the world.

Another method of communicating is via video teleconferencing, similar to Skype or FaceTime, using the same laptops and Ku-band system. NASA scheduled a twenty-minute call every weekend with our families, which required special equipment in my house to tie into the NASA system, so it wasn't quite as easy as FaceTime. But it was still nice to be able to see faces. Those weekend sessions were normally extended to an hour or so if desired. There was also a perk to being an astronaut: NASA set up two special video sessions with anyone we wanted—a celebrity or politician or musician, anyone at all. The NASA family support team would cold-call that person and say that an astronaut in space wanted to talk. They almost always said yes! My awesome family support helper, Beth Turner, set up some amazing talks for me, which were huge morale boosts in the middle of a very long cruise. We keep those events private for many reasons, but I can say for sure that you've heard of some of the folks I talked with. Those calls were a blast for all involved!

I often thought of the tribulations my grandparents went through in World War II—they left home and then had no contact for years, other than letters that were censored by their commanding officer. And now there I was, in outer space, not even on Earth, emailing every day. Making phone calls every day. Video conferencing weekly. There are many hardships that go along with spaceflight, but staying connected is generally not one of them. Today's astronauts are very lucky to be able to maintain contact with Earth, or cursed, depending on how you look at it.

HEARING VOICES

How Psychologists Prepare You
for What Spaceflight Does to Your Head

t is a fairly easy thing to hire astronauts who are technically capable. There are a lot of people out there who can repair broken equipment, perform spacewalks, fly complicated vehicles, or do science experiments. Those things are all important and require a wide range of technical skills, but at the end of the day there are a lot of technically skilled folks in the world. From my perspective, the much more difficult skill for an astronaut candidate was being psychologically adept for space travel. When I was reviewing astronaut applications, I would always look for clues to their suitability from a personality and psychological viewpoint, more than their technical ability.

Everyone who makes it through the first few gates of NASA's astronaut application process has serious technical qualifications. They will be a fighter pilot/test pilot, a medical doctor, or an accomplished scientist or engineer, along with having a broad range of experience in flying, climbing mountains, working on race cars, etc. We received more than 18,000 applicants for the last astronaut class, so it wasn't hard to find candidates with strong backgrounds. But the hard part was determining who would be psychologically suitable. Who would be able to handle the stress of constantly putting their life on the line, knowing that if the rocket malfunctions there is probably nothing they can do about it and they would likely die, or if a meteor comes flying through the station or their spacesuit at the wrong time, again they would probably die? Who could handle being stuck with a handful of other astronauts and cosmonauts for months at a time in a very confined space? For me, the soft skills were much more important than the technical skills when selecting new astronauts.

Given the importance of the psychological aspect of spaceflight, NASA designated psychologists to monitor each crew throughout their mission. We also had a family support person for each spaceflight, who was absolutely indispensable. Mine was a wonderful lady named Beth Turner. Beth would help my family set up their weekly video teleconference sessions with me, stay in touch with them throughout the mission to make sure everything was going well, and organize care packages to come up on occasional resupply ships, including the most critical item in space—chocolate. I just can't say enough good things about Beth; she was wonderful. For all of the formal psychological studies about astronauts' well-being, having a good care package, knowing that your family member is taken care of on their birthday, or getting a video greeting from friends was infinitely more valuable. Thank you, Beth, for keeping me sane.

I took part in an experiment called Journals, something that most ISS astronauts have done since the first crew launched in 2000. It is run by a gentleman named Dr. Jack Stuster, one of the top researchers in the world in the field of expeditionary behavior. He has written extensively about expeditions to the Arctic, as well as space. Most important, he's a good guy. During my time in space I kept a weekly journal, in the form of an audio recording of what was happening—what was fun, amazing, angering, frustrating, lonely, wonderful. . . . Whatever emotions I was experiencing I would chronicle, in audio format, for him to analyze. He heard my unfiltered thoughts and feelings more than anyone else during my time in space. Doing an audio recording was more convenient than writing—I could talk faster than I could type, so that worked well.

When I got back from space, I spent some time with Jack discussing the results of his analysis, and they were interesting. He was acutely aware of the frustrations I was facing, the joys I experienced, and the interactions with my crewmates. He knew what went well and what was frustrating. He knew these things because I was honest with my diary. I wrote, or more accurately narrated, exactly what was happening and how I felt about it.

It was a much different story with the official NASA psychologists. If your flight surgeon or official psychologist was alerted to the fact that something was wrong, there was always a possibility that management would be brought into the loop and they would take action, removing you from a spacewalk, modifying your schedule, or some other kind of official action. For this reason, pilots in the Air Force and astronauts at NASA are all the same—life is good when they talk to the doc. We used to say that nothing good could come of a visit to the flight surgeon—the best you could hope for was no change in your flight status, so I was always happy with the official psychologists. As a fighter pilot, I was used to compartmentalizing. Whatever was happening at home in my personal life, I left at home. I went to work and focused on flying the jet. The F-16 was very demanding; any lapse of concentration could easily kill you, so whenever I was in the cockpit I was focused 100 percent on flying the jet.

> Hollywood pilots and astronauts are always yelling and emotional. In real life, we are very staid and stoic.

That compartmentalization skill transferred very nicely to my career as an astronaut. Whatever was happening at home or at work, I was able to focus on my job. Cool. Calm. Collected. A prototypical test pilot. Flying jets, space shuttles, or space stations, I was always able to disconnect my immediate duties from what was happening in the rest of my life. When my wife was treated very badly by NASA management, I was able to remain professional and do my duties in space, leading my crew without hinting of problems on Earth. When there were real personal problems back home, I did my periodic PPCs (personal psychological conferences) with the NASA shrink without issue. Those brief meetings were actually enjoyable; we would joke and talk about sports and the weather, while fairly serious issues brewing in my life went undiscussed. This was entirely on me—I don't blame them for not picking up on those things, because if a patient isn't going to discuss something the doctor probably can't know about it. And in the end, compartmentalizing issues wasn't fair to my family either. But it was my way of dealing with problems—in the end, I didn't deal with them.

Hollywood pilots and astronauts are always yelling and emotional. In real life, we are very staid and stoic. When a flock of birds goes down the left engine of your jet and causes the compressor blades to explode on takeoff, and the fire light comes on and the engine seizes, you don't want your pilot yelling and screaming. The Air Force would occasionally bring a psychologist to our F-16 squadron to give us a briefing about the importance of opening up and not compartmentalizing too much. You can imagine how that went over to a group of testosterone-filled, overachieving fighter pilots. We nodded and moved on, heading back to the squadron to talk about the latest dogfighting maneuvers or the next deployment we were going on. Leaving personal problems at home. Laughing and joking and being pretty much the best pilots on the planet, at least in our own minds.

One of the best examples of how a crew's psychology can be affected involved a series of incidents with our cargo resupply ships. Just before my Expedition 42/43 launch, an Orbital Cygnus vehicle exploded on liftoff, and with it tons of supplies. A few months later, a Russian Progress ship exploded. A few months after that, a SpaceX Dragon exploded. Three lost resupply vehicles in an eight-month window. It was a challenging time for the ISS program, to say the least. But it also had an impact on us, the crew. In addition to the fact that our care packages were lost with each mishap, when the Progress exploded the Russians delayed our replacement crew, because they would be launching on the same kind of rocket as the Progress. That meant that they needed to do their investigation before the next crew would launch, which therefore meant that our mission would be extended in order to keep six people on the station, and not drop down to only three people for an extended period of time.

We were told only that we were extended, but not for how long—neither NASA nor the Russian space agency told us the duration. We were stranded in space for an undetermined period of time. Something like that could really stress out a crew psychologically, but I was proud of my crew, Samantha and Anton. They both handled this situation professionally and with a great attitude. I was the ISS commander at that time and held a daily crew meeting to talk about the delay—what rumors had anyone heard that day? Usually, there was nothing from the NASA side, because the Soyuz wasn't an American vehicle and we didn't have much to say. Although I received no updates from my boss in either the astronaut office or FOD (the Flight Operations Directorate, the next layer up in the NASA bureaucracy), the station program manager, Mike Suffredini, told me what he knew and tried to keep us in the loop with whatever information he had.

The best and most interesting gossip came from the Russian side. There were all kinds of theories for the cause of the crash floating around out there, from simple mechanical problems to conspiracy theories to even aliens. We always got a laugh from those crazy ideas, and it was good for the whole crew to get together and share what they had heard to ensure there was no rumor or discontent brewing on board. The bottom line was that we were stuck in space for an undetermined amount of time, and I wanted everyone to feel like there was an open flow of information. I made it very clear to NASA management that our crew was in good spirits—we would be ready to return to Earth that week, or in a month, or even an extra six months. I was so thankful to Anton and Samantha for having such a great attitude. As a crew, we used humor to cope. I began a March Madness–style bracket for the whole crew to guess what the delay would be, the bet being a bottle of the winner's choosing upon return to Earth. There were two axes in that bracket—the first was how long our return would be delayed, and the second was the delay of our replacement crew. The winner was paid in full during our debrief back on Earth, a few months later.

Personally, the key to my sanity during that uncertain time was to have a good attitude. I simply thought, "I'll have the rest of my life back on Earth, so enjoy it while you have some time in space." I'm a photographer at heart, and I had a lot of photos I still wanted to take, so I took advantage of those additional weeks to get those shots. I got some great southern lights photos, as well as an amazing shot of the pyramids on the 199th day of our mission (nothing like waiting until the last minute). I also was able to shoot more of the IMAX film *A Beautiful Planet* during that bonus month.

There have been other ISS incidents over the years that have delayed crews' returns, and it was usually a real bummer for the astronauts stuck in space. For me and my crewmates, that wasn't the case. The reason was our attitude. We were determined to make the most out of our time, focusing on the positive and realizing that the mission would come to an end soon enough, so we should enjoy the time in space while we could.

I think that attitude is the key to many of our situations in life. Make the most out of your circumstances. Enjoy what you can. Learn from what you can. Suffer through what you must. And learn from it. What doesn't kill you should make you better. If you go through life with that attitude, you will be happier and more successful than by complaining.

PACKAGE DELIVERIES

Receiving, Unpacking, and Repacking Cargo Ships

Space is a harsh place. That may be the understatement of the century. Although we have explored quite a bit of our solar system and are just now learning of the vast number of planets out there orbiting other stars, we have not found anywhere else in the universe where humans could live without artificial help. Earth is plan A, and there is no plan B.

The ISS is a stark illustration of this. The only thing homemade in space is electricity, from the solar panels. Other than that, everything must come from Earth: Food. Water. Underwear. Spare equipment. Uplinked *Game of Thrones* seasons. Life in space is entirely dependent on our home planet.

The mechanics of this resupply effort is a complicated affair. It begins at the home ports of our various cargo ships: America, Russia, and Japan all currently send vehicles to the ISS. Europe previously did, but they sent their final cargo resupply ship in 2015. Each of the partner space agencies tracks what supplies are needed, how much spare margin is required for food, clothing, etc., and when to schedule launches. This work is particularly vexing because there is so much to deconflict. First, you have to have a free docking port, because there are only so many hatches on the station and they are often taken up by visiting capsules. Cargo ships also cannot arrive or depart on spacewalk days, or on days when crews are arriving from or departing for Earth. Finally, there is an obscure orbital mechanics parameter called Beta that disqualifies a few weeks per year because during that time the station is overheated by the sun angle.

Once a good launch window is found, the vehicle needs to be loaded with supplies. Typical manifests include water (normally on the Russian Progress cargo vehicle), clothes, science experiments (sometimes even live critters), food, spare parts, and care packages for the crews. As for those critters, flying live plants or animals into space poses some serious challenges. You can put your underwear in the capsule a year in advance, and when it gets to space everything will be fine. But mice, or even plants, have a very limited window in which they can be loaded, and once they are loaded the clock is ticking. If launch slips, which happens all the time, they have to be unloaded, and a new "crew" loaded days later. I was shocked to learn that a typical crew of forty mice bound for space requires more than 1,000 to be ready on the ground. This allows a spare crew of mice to launch if there was a delay and also maintains a control group that scientists study simultaneously on Earth. A good friend and longtime NASA scientist assures me that flying animals into space is much more difficult than flying people, and based on all that she's told me about this process, I believe it.

Launch scheduled. Manifest determined. Supplies loaded. Plants and animals ready to go. Now it's time to light the candle and get the vehicle on the way to the station. Each cargo ship flies a different profile; the Russian Progress can dock within a few hours of launch, though other vehicles may take many days to get to the ISS. In all cases they fly autonomously by onboard computers, with their respective control centers watching over them. They all launch in the same orbital plane as the station, which means both the station and the visiting vehicle fly the same ground track over the surface of the Earth, and the cargo ships launch below and behind the station. Because they are in a lower orbit, they fly faster than the station, slowly closing on their target, orbit by orbit.

The big difference between vehicles occurs upon arrival at the ISS. The Russian Progress vehicle flies itself all the way to docking. There is a possibility of the crew taking over manually in the event that something goes wrong, but normally everything is 100 percent automated until the Progress is firmly attached to one of the Russian docking hatches.

The Japanese HTV, SpaceX Dragon, and Northrop Grumman Cygnus all fly up to a point about 30 feet below the station, where they hover, flying 17,500 mph in close formation with the ISS. At this point, the crew manually flies the station's robotic arm to reach out and grab the floating cargo ship.

This is done from the Cupola, the seven-windowed observation module that is awesome not only for taking pictures of the universe, but also for grappling visiting cargo vehicles. The view of another spaceship flying alongside your spaceship, with the Earth and universe in the background, is beyond words. After the crew captures the free-floating vehicle, Houston takes over control of the robotic arm and moves the ship to one of the station's docking ports, where it is berthed and firmly attached to the ISS using electrically driven bolts.

You can't just start unloading the 5,000 pounds or more of stuff without a plan.

At this point, the crew gets involved again. There is a tedious but important series of leak checks that are performed to make sure there is a tight seal around the hatch. We use an elaborate set of vacuum hoses, pressure probes, and the old-fashioned stopwatch function on our Omega watches to help us verify pressure integrity.

Finally, it's time to open the hatch. I had several opportunities to open hatches on new vehicles, and there is definitely a unique smell of space. It's hard to describe. But after all, what does a strawberry smell like? Or a new car? Or a wet dog? The best I can do is this: It's a bit like an electrical smell, ozone or sparks, kind of like my grandfather's old electric train sets. It's a little musty, but very mechanical. The bottom line is that it's a unique smell. I assume it's from the harsh vacuum and temperatures that the equipment has just experienced. It's tough to describe, but if I smelled it again I would recognize it immediately.

Next it's time to get to the cargo. This is where every crew would love to have an accountant on board, because unpacking and then packing a cargo ship is a *very* tedious and detail-oriented process. You can't just start unloading the 5,000 pounds or more of stuff without a plan. Each unpack plan had an overview of the order to move equipment as well as a detailed, line-by-line spreadsheet of every single item on the vehicle, a location to find it, and its final storage location on the ISS. We kept a printed copy of the unpack plan velcroed to the ceiling above the cargo ship's docking port, and whoever was doing the unpacking that day would be responsible for reading through the plan and carefully checking off each item unpacked. At the end of every workday, we would tag up with a storage specialist back on the ground and update them on our progress. A typical call would be, "We're complete through line 43, with the exception of line 22, and in line 33 the serial number was 1007."

Most cargo ships remain docked to the ISS for a month or two and then are loaded up with trash or other items for return to Earth. The process is essentially the same, only in reverse. Most of the vehicles burn up when they reenter the atmosphere in a fiery trail of station trash—we often joked of poor earthlings seeing a poop meteor streaking high above. In reality, though, these vehicles deorbit to a remote location in the Pacific Ocean, so there are rarely any humans on Earth who actually witness the last fiery gasps of our cargo vehicles. But the SpaceX Dragon has parachutes and a heat shield and is able to safely return cargo to Earth, so it is usually filled with science experiments, mice, equipment that could be refurbished and reused, and miscellaneous crew personal items that they wanted to return to Earth. We would often use our old T-shirts and clothes as packing material, and occasionally get a bag of those souvenirs on our desks a few months after they got back to Earth.

Attention to tedious detail has never been my strength. To illustrate this, one day I was unpacking some maintenance equipment from a SpaceX Dragon, and I unloaded all the items and put them where I thought they should go on the ISS. Except a few hours later, when my crewmates picked up where I had left off, one piece was missing. I realized I had mistakenly put an item in the wrong place, and on the ISS that basically meant it was lost forever. This was bad, because we had a spacewalk coming up and we would need the missing hardware. The whole crew got involved, and after several hours the missing part was found, and boy did I learn my lesson. Don't lose stuff—do everything necessary, pay attention to detail, be OCD when tracking it, but just don't lose stuff!

Before my spaceflights I never thought much about the business of resupply, and I assumed that it might take a day or two to unpack a load of cargo. But this cadence—reading the plan, finding the item, moving it to its location, checking it off the list, and then repeating—defined the rhythm of life on the station for several weeks or more every time a vehicle showed up.

This made me think of how in the Air Force all the glory went to the fighter pilots, but having a refueling tanker was absolutely essential for us fighter pilots to do our mission. Those tanker guys had a saying, "Nobody kicks ass until the tanker passes gas." In the same vein, launch and rendezvous and spacewalks are where the glory is, but at the end of the day, if the cargo ships don't show up with gear and if the crew doesn't properly unpack and stow this precious cargo, spaceflight wouldn't be possible. And while fighter pilots may rule in the Air Force, maybe on the space station accountants should rule!

NETFLIX, HULU, AND BASEBALL

In—Flight Entertainment

rarely watch TV on Earth. I'll occasionally watch a news program or ESPN, but in general I just don't watch normal television shows. In fact, I just recently saw *Friends* and *The Office* for the first time. I didn't even know the main characters until my kids explained them to me. So when I got to space, I wasn't too terribly disappointed to not have normal cable TV. Little did I know that watching TV and movies and other programs would be one of my favorite and most relaxing activities to help maintain my sanity while off the planet.

One of the most important things to do before flight is to pick out music and video entertainment. The mechanics of getting that entertainment to astronauts in space are interesting. In years gone by, astronauts would bring up CDs or DVDs. I brought a collection of these on STS-130, but there was no time to use them. Beth Turner was my psychological support person for my long-duration mission and she facilitated this process. I gave her my Pandora password and she recorded my stations onto three-hour-long MP3 files, eventually uplinking seventy-six of them, from every genre—rock, pop, dance, Christian, country, chill, jazz, classical, alternative, etc. Those music files kept me sane; I would listen to them via a portable Bose speaker throughout the workday, hoping Houston wasn't watching me lip-sync.

I also listened to other programs, my absolute favorite being *Car Talk*, which ran in various forms on NPR from 1977 to 2012. It was one of the most beloved radio talk show programs of all time, hosted by Tom and Ray Magliozzi, known also as "Click and Clack, the Tappet Brothers." The brothers liked to brag that they won NPR's "most likely to be canceled" award for twenty-five years running. I listened while I worked out, and my crewmates would often float by and wonder why I was inexplicably laughing to myself. I suppose they thought I was crazy, but those *Car Talk* guys were funny and they made my spaceflight a little more fun.

Besides audio, we had quite a few video options. The ground would uplink *NBC Nightly News* every day, as well as occasionally livestream either CNN or ESPN. Beth also let me pick out TV programs and movies, which were amazingly wonderful to have while I was working out. I asked for the TV show *24*. It was a decade old, but I had never seen it, and I made it through four seasons while in space. Go Jack Bauer! I also got caught up in *The Americans*, a KGB/FBI spy thriller series. Beth also sent up clips of my Orioles and Astros, starting with spring training. Spring of 2015 was a good time to be an Orioles and Astros fan—both teams were good and fun to watch. *The Middle* was another of my favorite TV programs. In fact, I did a video Skype one day with the cast of the show from space, which was a highlight for me and I think for them as well.

Movies were also a great way to relax. There is a digital library of movies in MP4 format on board the ISS, with maybe a few hundred titles. Beth sent up a few requests for me. *Interstellar* was one of my favorites, not because it was a space movie, but because it was about a father and daughter, and I was missing my daughter. A friend recommended some eclectic movies, including *The Lives of Others*, a German-language movie about the Stasi and 1984 East Berlin, which was quite good. *In Bruges* was another foreign film I enjoyed. Watching *2001*, *Gravity*, and *Avatar* in space was a little surreal. I even watched *Groundhog Day* on Groundhog Day, with Samantha Cristoforetti, as a bit of American culture training. I think she got the point after that scene when Bill Murray died and then unexpectedly woke up yet again to "I Got You Babe" (we didn't finish the film, in case you're wondering). Interestingly, the Russian psychologists uplinked *50 Shades of Grey*, in Russian. The first time I had watched *Interstellar* it was in Russian and I really didn't understand much. However, with *50 Shades* subtitles weren't required.

The Russian psychologists did something else that was a *huge* hit with the whole crew. One day I was floating through Node 1 (the center of the ISS) and I heard a bird chirping in the Node 3 module, where Misha Kornienko was exercising. I asked him where the bird was and he let out a big, deep laugh. He explained that his support team had uplinked sounds from Earth—jungle birds, café noises, waves crashing on the shore, rain. Those sounds were amazing! I had been in space for about four months by that point and I had no idea how much I missed them. In fact, the whole crew loved them so much that we played the rain sound on all of the station's laptops one weekend. It was great on Saturday morning, but by the time Sunday came around we were ready to jump out of the airlock, so we went around and turned the rain noise off. It impressed me how powerful the sounds of our home planet were, lifting our spirits. As I mentioned before, I often used these sounds to drift off to sleep at bedtime.

I never would have thought that music and TV and movies would be such an important aspect of spaceflight, but they really played a huge role in keeping my morale up. Thank goodness for digital uplinks. And most of all, thank goodness for Beth Turner!

FIGHTER PILOT DOES SCIENCE

Experiments Are the Real Point of the Mission

The mission of the space station is science. That's why a coalition of nations built this amazing castle in the sky. It is why the ISS program has been funded by congresses, dumas, and parliaments, for billions of dollars per year. It is why astronauts and cosmonauts have risked their lives since the first station construction mission back in 1998. And it is why the ISS has a bright and long future ahead of it. Science.

Because I came to NASA as a fighter pilot, science was not a proper part of my professional background. I had taken physics, biology, and chemistry in college, and was a bit of an amateur astronomer, avidly reading *Astronomy* and *Sky & Telescope* magazines as a kid. I especially enjoyed CMO (Crew Medical Officer) training.

I loved all of the science experiments I did. On both my shuttle and station flights, they were a highlight of my astronaut career. I was always aware that for each experiment there was a PI (principal investigator) and a team of PhDs and technicians back on Earth who had spent years of their career working on it, while for me, most experiments took only thirty minutes to perform, and then I'd be off to my next task. So I always tried to take my time and pay each experiment the attention it was due.

There are many different types of experiments that are done in space, ranging from simple boxes containing scientific equipment that are left alone with no astronaut interaction to very intensive experiments that require detailed crew work for extended periods of time. During my 200-day mission there were more than 250 experiments conducted on the ISS, from just about every academic discipline you can imagine: physics, biology, chemistry, engineering, medicine, astronomy, psychology, materials science, etc. If you can take it as a class in college, there is probably an experiment on the space station in that field. Most NASA science is controlled from the payload operations center at the Marshall Spaceflight Center in Alabama, as opposed to mission control in Houston, which is responsible for controlling the vehicle itself.

The most interesting were experiments that required my active participation. For example, CFE (Capillary Flow Experiment) was a canister with a silica-based gel in one end, and as a vane rotated within that canister, the fluid would suddenly flow up to the other end. Imagine rotating a spatula in a bowl of brownie batter, and as the spatula rotated 90 degrees in the bowl suddenly the batter would flow up to the top of the handle. Weird, but that's what happens in space, by a mechanism called capillary flow. I had the opportunity to do this experiment several times; I set up a series of video cameras to capture the motion and precisely measured the angle of rotation of the vane within half-degree increments, noting the value when the fluid began motion. This was a potentially useful investigation, because capillary flow might eventually benefit satellite manufacturers as they try to get every drop of fuel out of their gas tanks. I most enjoyed it because I was actively involved.

A very different investigation was MICRO-5, a microbiology experiment that involved infecting *C. elegans* worms with salmonella and *E. coli*, with the hope of finding better vaccines for those diseases. I spent several full days in the glovebox, a dryer-size piece of hardware that allowed us to put our hands in and work on just about anything, while keeping it contained behind glass walls. Besides disease-laden worms, we worked on 3D printers, fragile materials, mice, and myriad other things that you wouldn't want floating loose in the cabin, filtering into your lungs, or getting stuck in your eyeball.

One Saturday I spent the whole day working on those poor worms. A week prior they had been active, wiggling around in zero g until I introduced salmonella and *E. coli* into their diet. By Saturday they were wiggling much more slowly because they were feeling under the weather. My task was to put them into the MELFI (NASA acronym for freezer), rapidly freezing them in bulky black metal bricks the size of ice cream sandwiches that had been prechilled to –95°C. Scientists on Earth would thaw them months later, after they returned to the planet. Getting all twenty samples frozen and stored in the MELFI at the same time was challenging, to say the least. It looked like a rock concert as ultra-cold smoke, or condensation, poured out of the MELFI. It seemed as though every time I put one black brick into the freezer, two would float out in a cloud of white fog and black metal containers. Nailing jelly to a wall would have been easier, though finally I managed to wrangle everything into its proper place, or so I thought.

The next day I was floating through the US lab when I noticed something move out of the corner of my eye. Sure enough, there was a black container of disease-infected worms floating up in the ceiling of the lab. I was mortified. One of those suckers had apparently floated away during the herding, and the science from that particular sample must have been lost. Worse than that, there was a vial of either *E. coli* or salmonella floating loose in the lab, by now comfortably warmed to room temperature. I had a quick meeting with my commander, explaining the mistake I had made. His advice was to just put it back in the freezer and not tell anyone. I didn't want to do that because the scientists on Earth might then get false data and come to a false conclusion from this experiment. I'm no biologist, but I assumed having a worm thaw out and then refreeze might affect their data. So I called Huntsville and fessed up. They were understanding; they noted the serial number of the thawed worm container and had me put it back in the freezer. I was careful this time to make sure nothing else accidentally floated away.

A few months later, I had a chance to talk with those scientists during our debrief at the Marshall Space Flight Center in Huntsville, Alabama, and they were very happy with the results they had gotten from the experiment. I learned an important lesson from the worm incident: If you make a mistake, fess up. It might prevent others from making the same mistake in the future, and in the case of science, it ensures accurate results.

One of the more unusual tasks I had in space was to launch satellites. There are several different flavors of micro satellites that are launched from the ISS, and the first one I launched was called SPINSAT. It was a large sphere, about twice the size of a basketball, and as you may have guessed, it spun. I always wondered what they would call a spherical satellite that didn't spin—Wakefield-SAT? Once prepared for launch, it was attached to a movable tray inside a small airlock in the Japanese module. We shut the hatch between the airlock and the interior of the station, pumped the air back into the ISS, and opened the outer hatch that led to space. The tray then slowly moved outside, carrying the satellite with it. Next, the JEM-RMS (Japanese robotic arm) grabbed the satellite and moved it into position to hand off to the SSRMS, also called Canadarm2 (the big ISS robotic arm made in Canada, the first having flown on the space shuttle), which grabbed the satellite and slowly moved it into release position, below the station. A few hours later, we all crammed in the Cupola to watch and photograph the big event. We released it after a brief countdown and orbital mechanics took over, slowly pulling the SPINSAT below and then in front of the ISS.

I performed this task several times, later on CUBESATs. You guessed it again, these were cube-shaped satellites, about the size of a small cooler. But the procedure was the same: Prep the satellites in their quad-launcher box, attach it to the JAL (Japanese airlock), close the first hatch, suck as much air back into the ISS as possible to prevent waste, open the outer hatch, grab it first with the Japanese robotic arm, hand over to the SSRMS, move into position, and cram into the Cupola to watch the show. I think it's safe to say this is a favorite task for ISS astronauts.

Not all experiments worked. ROBONAUT (surprisingly enough, an astronaut-like robot) was popular from a PR point of view. It was cool to show this modern/retro robot, who looked like a cross between Robocop and a Terminator (the bad kind, from *Terminator 2: Judgment Day*). ROBONAUT had just gotten his legs and was about ten feet tall when fully extended, which was a little intimidating. Early in my mission, I was tasked with dragging him out of his refrigerator-size storage closet, unfolding and powering him up, and waiting while the ground engineers ran software tests. That was a mentally grueling day for me; I was very slow finding all of the connections and manually rotating all of his arm and leg joints to the proper configuration. Once everything was finally set up, it was time for the big moment. Power switch—on. Nothing. Recheck all the cables and cords. Power switch—on, once again. Again, nothing. I went through troubleshooting for a long time, but unfortunately, it just didn't work.

I had a similar issue with CSLM (a materials science experiment), which was performed in the glovebox. I would set up the equipment and it didn't work. No matter how much troubleshooting I did, it just wouldn't power up. It turned out that particular experiment was past its planned life; they had already gotten all of the data they had planned and were just running additional bonus tests because there was some free time in my schedule. However, I didn't know that, and I thought I was doing something wrong that was causing it to not work. In the same way, ROBONAUT had a known issue with the software and the ground wasn't surprised that it wasn't working, but I thought it was all my fault. Incidents like these taught me another lesson: If something doesn't work, do some quick troubleshooting passes and then tell someone. There may be a known condition that is causing the problem, and it's better to know that right up front rather than wasting a lot of time and emotional energy feeling badly about something that may not be your fault.

Later I learned that NASA had a new initiative to get experiments to the ISS faster and cheaper. This meant that they would fail at a higher rate, but that was OK in NASA's eyes. They would rather have 100 experiments with a high failure rate and low cost than fifty expensive ones that mostly worked. I liked that philosophy; I just wish I had known that before beating my head against the rack (ISS term for wall) time and again when experiments didn't work as planned.

There were many different experiments that I took part in. Some were memorable, some forgotten. Some interesting, some excruciating. Some valuable, many . . . less valuable. Most worked, some didn't. But doing science was the raison d'être for the ISS, and it was why we risked our lives riding a rocket into the vacuum of space. It was an enjoyable and memorable part of my missions, and I'm thankful to have had the chance to be the hands, eyes, and ears for those scientists who entrusted me with a significant phase of their careers. I'm also thankful that the black brick didn't leak salmonella into my eyeball.

MAROONED

What to Do If You're Stranded Up There

"Houston, Tranquility base here, the Eagle has landed. . . ." It's been more than fifty years since those words were uttered, and most everyone is familiar with the Apollo story and the brave crews who flew to the Moon and returned to the Earth. A lesser-known chapter of that story was a speech that was prepared for President Nixon, just in case Neil and Buzz had been stranded on the Moon. It was a beautiful text, thoughtful and well written by William Safire, presidential speechwriter. It began:

Fate has ordained that the men who went to the moon to explore in peace will stay on the moon to rest in peace . . .

Thankfully, Mr. Nixon never had to utter these words, and the Apollo astronauts safely returned from the Moon. But spaceflight is an inherently risky business. There have been three fatal NASA accidents (*Apollo 1*, *Challenger*, and *Columbia*) as well as two fatal Soyuz accidents in the Soviet Union. This danger raises a question that we hope and pray will never need to be answered, but one that is worth asking: What do you do if you're stuck in space?

When I was training to be an Air Force pilot, I had to learn about ejection seats and the basics of parachute landings. It turns out that a lot can go wrong with your parachute during an ejection: It can be stuck in the seat, once it deploys the lines can become tangled, parachute panels can blow out, causing a faster descent rate, or it can have a "Mae West," with a riser line wrapped over the parachute, causing it to have two rounded bulges instead of the normal round parachute shape. These problems all have fixes—you can pull yourself up to the chute via the risers, then let them go and hope that they will pop open the chute properly. You can manually deploy the chute if your chute is stuck in the backpack. You can delay the four-line jettison procedure that allows you to steer the chute, but that also causes it to sink faster. There are lots of things you can do if you have a parachute malfunction. But I'll never forget one important piece of advice an old and crusty sergeant gave me when I was going through pilot training at Williams AFB in Arizona, during the fall of 1989: "After an ejection, you have the rest of your life to get the parachute deployed."

That is such great advice for so many situations in life. Some problems simply must be solved or you can't go on. If your brakes stop working in your car. If the landing gear won't come down on an airliner. If you have a serious financial problem at work (you may not die, but the business might). Sometimes you need that sense of urgency—forget everything else and fix this problem, or it will be a very bad day.

This same philosophy would apply if you were stuck in space. Let's say, for example, that the rocket engine that is required to perform your deorbit burn wouldn't light. Or the heat shield was fatally damaged so that you couldn't survive the re-entry. Or the computers required to precisely guide the spacecraft through re-entry stopped working. There are a thousand scenarios that would cause you to be stuck in space, but the real question is, what would the astronauts do?

Does Houston tell the astronauts about a potentially fatal malfunction? Does the crew try to fix the problem? Or end it all?

The first scenario, a broken engine, is something that I trained for extensively, in both the shuttle and Soyuz. When you're orbiting Earth you need to slow down in order to bend your trajectory down to the atmosphere, because once that happens friction and drag take over and you will inevitably return to the Earth's surface. But unless you have a rocket engine that's functioning, you will be stuck in orbit for years, if not centuries, before the tenuous atmospheric drag at orbital altitudes will eventually bring you down. No spaceship has enough supplies to last that long, so you've got the rest of your life to figure out how to get the rocket engines working.

Typical tricks to restore function that astronauts use in simulated engine failures are cycling valves, computer systems, or electrical power supplies. If there's a propellant leak, the first goal is to stop the leak ASAP by closing valves. If that doesn't isolate the leak, you're hosed, unless you can do an immediate deorbit burn on the remaining propellant, which means landing wherever your current trajectory takes you. It would be on a random location on Earth, which isn't necessarily a great thing. Alternatively, if you have enough propellant left, you may able to return to the ISS before it all leaks out and then wait there for a rescue ship.

An interesting corollary to the broken engine in orbit is a broken engine while on the Moon or Mars. There may be a possibility of repair, but that's very unlikely. And unless there were another lander on the planet with you, or not too far away, you are also hosed. It comes down to how much oxygen you have and when the next vehicle arrives from Earth. The answers to those questions are probably not what you want to hear. Which would lead to everyone's fear—stranded there, fully conscious of what's going on, and no way to escape. You'd either have to wait around until your oxygen supply ran out or end it deliberately. I'm sometimes asked if we carried poison pills for a scenario like that. I never heard of any suicide pills. But when you're in space, opening the hatch would take care of that problem very quickly. It's a scenario that we all hope never comes to pass, but it was one that even President Nixon was prepared for.

On STS-130 we spent the better part of two days inspecting *Endeavour*'s fragile heat shield. That was because of the painful lesson we learned after STS-107, when *Columbia* was destroyed after having her heat shield fatally damaged from foam debris during launch. This is because a spaceship just can't survive heat that reaches 4,000 degrees without an intact shield. So if your heat shield is damaged, and you know about it before you leave the station, you have two options: Wait there for a rescue vehicle or try to repair it. There is no way to repair a capsule heat shield, but on the space shuttle we had an elaborate system of spacewalk repair techniques—foam gel to fill in damaged thermal tiles on the bottom of the orbiter and plates to cover holes in the critical leading edge of the wing. Those techniques were practiced on test spacewalks but thankfully were never used on the actual heat shield of an orbiter.

On STS-107 a piece of foam popped off the fuel tank into 500-knot wind and hit the leading edge of the wing with incredible force, making a large hole in the critical composite thermal protection system. NASA knew about the foam strike and assumed that there was some damage to the heat shield, but after twenty years of flying the shuttle with only harmless foam damage, management decided not to take pictures of *Columbia*'s wing. They also assumed that there was nothing that could be done if there were serious damage.

This was insane in my view; as an astronaut I would always want to know the status of my vehicle. And there was a shuttle on the launchpad, about a month away from being able to launch on an impromptu rescue mission. That mission would have been untested and dangerous. The *Columbia* crew would have had to turn off all equipment and spend a miserable month in the small crew area waiting for rescue. Another option would have been to do an impromptu spacewalk to try to repair the hole, which would have almost certainly been unsuccessful. But the alternative to trying these unlikely-to-succeed options was that we lost the crew. I saw the video of the foam strike a few days into their mission and brought it up with my management, and I was assured that it wasn't a "safety-of-flight issue." The biggest regret of my career was that I didn't push the matter further, but as an unflown new guy, I just took their word for it.

Although both the *Challenger* and *Columbia* accidents had very technical explanations, at the end of the day they were both management mistakes, made by some of the smartest people on Earth, who were very motivated for mission success. Those mistakes were not evil or intentional, but they serve as a cautionary tale of how even the best-intentioned and smartest leaders can be blinded to reality. I hope and pray we never have to do another accident investigation. But the reality is, spaceflight is dangerous and unforgiving. And sometimes you need to try something untested and bold in order to survive the harshest environment imaginable. Just ask Gene Kranz and his *Apollo 13* team. You know, they should make a movie about that story. . . .

SPACEWALKING

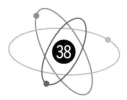

THE WORLD'S BIGGEST POOL

Training Underwater for Spacewalks

They say the Neutral Buoyancy Laboratory (NBL) is the world's largest pool. I don't know if they've gone around the world and measured all the other pools, but one thing is for certain: It is one big pool. It needs to be big so it can contain a mockup of the International Space Station, which itself is bigger than a football field. In years past, it also contained a simulated space shuttle and *Hubble Space Telescope*, and it currently houses an Orion capsule and even a helicopter trainer for Gulf of Mexico oil rig companies. But the star of the show is clearly the ISS, submerged in 40 feet of water. The underwater simulated ISS is only about half the size of the real thing; the vast majority of the Russian segment and about half of the station truss structure are missing because of size and expense constraints, but the modules there are exactly what we need for training.

Astronauts from most of the world's space agencies have been using the NBL to learn how to perform spacewalks since it first opened in 1996. Neutral buoyancy spacewalk training dates back to the first American to perform a spacewalk, Ed White, who had tremendous difficulty moving around and performing his tasks in 1965. After that arduous EVA (extravehicular activity, the NASA acronym for spacewalk), Buzz Aldrin came up with the idea of doing training underwater to learn to float and work in outer space. That idea enabled the last fifty-plus years of spacewalking.

The underwater environment is different from space in several key ways. First, you're not weightless; you actually hang in your EMU (NASA acronym for spacesuit, or Extravehicular Mobility Unit). Six-hour NBL runs always gave me two big bruises on my chest, where the spacesuit metal rings dug into my skin. Second, there is a righting moment, which means the water tends to flip you heads-up, making it difficult to stay on your side or upside down in the pool. Finally, it's hard to move—*really* hard. You have to displace 400 pounds of water every time you want to move from point A to point B, and that requires a lot of strength, pushing constantly against the water. In order to stop you simply stop pushing, and the water stops you in a few seconds. In space it's the opposite; the slightest pressure and you start moving, but you have to work to stop when you get to your destination. This leads to a bit of negative training, because the NBL encourages strong, continuous force to move around, but in space you need to move slowly and gently. It's not too onerous, and bad habits can be quickly unlearned while in space with some mental concentration.

The two sad astronauts in their fancy space underwear would sit in the briefing room while twenty trainers, safety divers, and technicians discussed the day's plan, safety measures, specific tools and procedures, etc.

The first step in spacewalk training is to learn your hardware. There are seemingly millions of pieces of equipment, most complicated, all named with an acronym or nickname that you've never heard before. After learning the equipment nomenclature, it's time for a "one g" session, meaning tasks are performed in one g, on Earth, as opposed to zero g when you get to space. This occurs a few days before the actual underwater run, and the instructor reviews the tasks, procedures, and specific hardware that you will use. Next comes a scuba dive, where you and your spacewalking partner run through each task in normal scuba gear, before the big run in the EMU.

As a new astronaut I found scuba sessions to be invaluable to get familiar with the work sites, to learn translation paths, and also to learn the best body position for our specific tasks. But when I was assigned to an actual spaceflight, I stopped them for one simple reason—I wouldn't be able to do a scuba run on the ISS when tasked with a real spacewalk. I would have only virtual-reality software and conference calls with instructors on the ground to prepare for a real EVA, and I needed to get comfortable preparing for spacewalks without scuba gear.

There is a very repeatable rhythm to game day. I always began at Shirley's Donuts & Kolaches at 0545. For the eight years that I did NBL training, the same group of old men were at that doughnut shop in suburban Texas having breakfast, reading the paper, talking about high school football, or debating politics. We'd smile and give a polite nod, and I'd get my dozen doughnuts and dozen kolaches (a Texas breakfast foodgasm of a warm roll enveloping meat and cheese) and head off to the "house of pain," my personal nickname for the NBL. Spacewalk practice was honestly about 96 percent pain and only a little bit fun. This breakfast junk food was my unofficial peace offering to the training team, as if to say, "Please go just a little bit easier on us today." I don't think it was ever successful, but the doughnuts were good.

Upon arrival at the NBL, I would go straight to set up the day's equipment, laid out poolside. It took about thirty minutes for equipment prep—the MWS (NASA acronym for the tool caddy attached to the chest of the EMU, which is the NASA acronym for spacesuit) needs to be configured with wire ties, RETs, waist tethers, PGT, trash bag, BRT, MWS-EE, AET, safety tether pack, etc. Plus configuring the specialized hardware for the task of the day: scoops, TMs, socket caddies, RPCMs, LEEs, ad infinitum. Did I mention you have to learn a new language of NASA acronyms before being allowed to spacewalk?

Then it was time to head to the locker room to put on the first layers of the spacesuit. Just like launch, first came the diaper. Next a set of moleskin for hands and knees to prevent skin damage, something unique to the EMU. Then a basic set of thin long underwear in addition to the LCVG, a big bulky long underwear garb full of plastic tubes to carry cooling water to keep your body from overheating, similar to the launch suit. Except this LCVG was much bulkier and had personalized pads sewn in it to protect your specific body type in the spacesuit. Those pads could protect your shoulders, knees, hips, or elbows, depending on how much of the suit you filled out. Next, blue booties over your feet. A final trip to the bathroom, a five-hour energy shot, a few ibuprofens, and off to the prebriefing. The two sad astronauts in their fancy space underwear would sit in the briefing room while twenty trainers, safety divers, and technicians discussed the day's plan, safety measures, specific tools and procedures, etc.

Two hours after arriving at the NBL, equipment ready, prebriefing over, it was finally time to suit up. First, the obligatory photos with whatever tour group or celebrity happened to be there that day. Next, the torture of squeezing into the EMU. I'm quite certain medieval British monarchs would have been *very* interested in the design of this spacesuit, because it surely would have given the iron maiden stiff competition. My upper body is big and not very flexible, which meant I had to hyperextend my elbows to squeeze into that thing. It would always leave bruises, but after a few minutes of wrestling and choice words I would finally wiggle in, my head popping through the neck ring, and I'd make the same joke every time: "It's a boy!" The best was yet to come, though: the helmet. As I said earlier, my head is huge and it required a painful technique to squeeze that helmet on.

Once in the approximately 400-pound suit, a special crane would gradually lower me and my wingman into the pool, and the moment of entering the water was always interesting. It immediately felt warm, as the suit squeezed my body from water pressure, and my vision was instantly blurred due to refraction. Our safety divers would then drag their two spacewalk trainees through the water from the crane to the airlock, adjusting our suits with a series of weights and buoyant floats that would attempt to keep us neutral in the water, not rising or sinking, and basically heads up. This was more art than science; the astronaut and diver had to work together to get this critically important weigh-out right. Even a small imbalance in buoyancy would require constant effort to stabilize yourself, which would quickly exhaust even the most physically fit astronaut. The divers also had to deal with the oxygen hose from a tank on the surface to the astronauts underwater. In space the suit managed O_2 and CO_2 on its own, but on Earth the hose ran 100 feet or more to the topside of the pool.

Weigh-out complete, the two astronauts (along with a team of three divers for each spacewalker) moved over to the ISS airlock to begin the run, where we were turned over to the instructors to begin the torture—I mean training. They talked us through each step, from egressing the airlock to moving to our work site on the simulated station exterior to getting our tools ready, etc. Everything was choreographed for maximum efficiency, because

A massive crane prepares to lift over 500 pounds of my body, spacesuit, and equipment into the NBL, our spacewalk training pool.

there is absolutely no time to waste on a real spacewalk. It's dangerous outside and there are a lot of potential problems that could cause a spacewalk to go south in a hurry, so we used the NBL to make the real-day spacewalk as efficient as possible.

Moving underwater was very different than moving in outer space. Underwater, it was extremely difficult to get moving and easy to stop, and you were always rotated to a heads-up orientation, none of which was true in space. Nonetheless, there were several invaluable lessons to be learned in the pool. First, the spacesuit is *really* uncomfortable and difficult to operate in. It's big and bulky and vision is limited. Second, keeping track of tethers and equipment is a constant chore. There was always a safety tether (a long, retractable wire) connecting you with the station. If it ever broke *and* you let go of the ISS, you would have a small SAFER (NASA acronym for jet pack) to fly back to safety. There were also multiple equipment tethers to prevent your stuff from floating away and local tethers that were used to keep you in place at a work site.

Keeping track of all of this was like herding cats while trying to hold on to a greased pig and counting backward from 100. In Russian. It's hard to do physically and even more difficult mentally. Spacewalking is a difficult skill, and the NBL does a great job of helping astronauts master it before going outside on the real day.

Another critical lesson was the importance of body position. If you could rotate yourself around to put the task right in front of your chest, you could probably get it done. If it was off to the side or above or below you, even the simplest task would quickly become impossible because of the difficulty and pain of moving your arms or reaching your hands in that bulky iron maiden (aka EMU). Worse than being difficult, reaching or stretching your arms, especially overhead, was downright dangerous. Every year I was at NASA, an astronaut or two would get shoulder surgery for a torn rotator cuff suffered during a training run at the NBL. Our frail shoulders were especially susceptible to damage when upside down in the pool, because your entire body weight would rest on your clavicle, pressing on metal rings in the suit. Raising your arms up over your head could potentially result in permanent damage to shoulder ligaments. The good news is that as long as they caught the injury while you were still an active astronaut, NASA would pay for the surgery. Other than that, you would have to file a workman's compensation claim and hope for the best. Injury potential, acute pain, and general discomfort were all good reasons to avoid going upside down in the pool if at all possible. Thankfully, this was not an issue in the weightlessness of space.

> Thanks to my fighter-pilot instincts to always sound cool on the radio, I calmly said, "Safety diver, you can take me back into the airlock."

During eight years of training at the NBL, I had some memorable runs. Several of them involved going into the truss of the station, an external structure made of pipes, bars, and wires that holds the solar arrays, which are attached more than a hundred feet from each side of the main part of the station. It was both fascinating and daunting to wedge myself and my 400-pound EMU inside that truss structure to work on something called an FHRC, a tombstone-looking, refrigerator-size contraption that allowed ammonia cooling fluid to pass from a stationary piece of the truss to a rotating piece. From the astronauts' point of view, this was just one of a thousand black box components that we would potentially have to install and replace during our mission. The internal workings of these devices were not our concern; we just had to remove the old and install the new. A unique challenge of ammonia hardware was that it required special steps to install and remove fluid lines, which was a big part of the pool training. Once the bad FHRC was removed, a process that would take an hour or more, I would hand it off to my poor crewmate, hopefully without banging into about a hundred things that were strategically located to be easily banged into. He would then store the broken device and retrieve a new one to install.

Learning to work as a team, with clear and concise communication between spacewalkers, an astronaut inside the ISS who was flying the robotic arm, and mission control in Houston was an important skill. Some may argue that communication was the most important as well as the most difficult skill we learned in the NBL. It may sound simple, but it's not, and until you have so many things going through your brain at once—keeping yourself attached to the ISS, keeping track of all of your equipment and tethers, keeping track of the schedule and pacing, keeping track of your crewmate, talking with Houston, all while being separated from instant death by just a 1-millimeter-thick plastic visor—you have not experienced *busy*.

Another memorable task I trained for in the NBL was for the LEE R&R (NASA acronym for removing and replacing the hand of the SSRMS, the NASA acronym for the Canadian robotic arm). The station's robotic arm has been in space since 2001, and because it's a mechanical system it occasionally breaks. So we put together an awesome team of fellow spacewalker Jeanette Epps and team lead Faruq Sabur to develop the procedures that would be used when the LEE eventually broke. It was satisfying to see our work put into use when the crew on the ISS replaced one of the LEEs in space a few years later. The road to that point ran through the NBL, where Jeanette and I reviewed procedures, worked with Faruq, and practiced them in the pool, in an iterative process that took about six months.

A big challenge for the LEE R&R was how to move that washing-machine-size thing. We used a technique called inchworming. I would say, "You have the LEE," Jeanette would acknowledge, I would move my body a foot or two, extend my hand, she would slowly float the LEE to me and say, "You have the LEE," I would acknowledge, she would then move herself, over and over. In this slow but deliberate manner, we moved that giant piece of broken equipment roughly 100 feet to its storage location. Then we'd grab the fresh LEE and inchworm it all the way back to the arm. It's slow, but it works. If we had accidentally let go and lost the multimillion-dollar LEE (Canadian dollars, eh?), we would have been given a nickname, so we were very careful when moving it.

A humorous story occurred on one of my first NBL runs, back in 2006. I was practicing egressing the space shuttle airlock, which is not easy. After trying a few different techniques, I came up with a brilliant idea; I would float out of the airlock on my back, looking up at the pool surface. This actually worked very well and I came right out of the hatch with no problems, but I just kept on going and going and going. I had accidentally let go of the handrail and shot myself directly into the shuttle's payload bay. In real life, I would have had a safety tether to slowly bring me back, but in the pool I just floated out, not holding on to anything. Not exactly an optimum way to begin my spacewalking career. Thanks to my fighter-pilot instincts to always sound cool on the radio, I calmly said, "Safety diver, you can take me back into the airlock." But man, did I give my instructor a lot to laugh about. On the positive side, I never again let go of structure. Some lessons are best learned by making a mistake!

In spite of nicknames like "house of pain," and failed attempts at bribery with doughnuts to minimize that pain, my training at the NBL was absolutely critical to my three spacewalks on the ISS. Some of the best people at NASA work in the EVA community, and I'm forever in their debt for the torture (actually I mean training) they put me through.

THE ART OF PUTTING ON A SPACESUIT

And You Thought Launch Was Complicated . . .

When you think of astronauts doing spacewalks, you think of them being outside, talking in cool astronaut language, with the beautiful Earth below, calmly working on their tasks and enjoying the view. However, well before any of that happens, there is a long and elaborate process of getting suited up. In NASA-speak, the two people who go outside are called EV1 and EV2 (extravehicular 1 and 2). The person on the inside who helps them get suited up is called IV (intravehicular). There was never any doubt in my mind, nor is there in most of my colleagues, that the job of IV is much harder than actually doing the spacewalk. The slightest error while suiting up the spacewalkers could very easily result in their death, so the stakes were as high as possible for Samantha as she got us suited up for the most extreme activity a human being can do.

The process of getting into an EMU (acronym for go-outside spacesuit) is much more involved than getting into a launch suit, and it takes much more preparation than it did for NBL training. In fact, the process begins many weeks before a planned spacewalk. First, the spacesuits are arranged and sized properly for each person. The HUT (acronym for upper-body section of the spacesuit) comes in a M, L, and XL, so the appropriate size is set aside for each of the two spacewalkers. Usually, there is one M, two Ls, and one XL available for use on the ISS, though that can change. If both spacewalkers want a size M or a size XL, one will have to bite the bullet and take a size L. It's not optimal, but it's a limitation we live with. Also, recently there has been a lot of press about the fact that there aren't more spacesuits for smaller crewmembers such as women.

A quick word about suit sizes. I have a big chest, and after I tried to squeeze into an L several times during years of training, the EVA team in Houston told me I was forbidden from even trying it again because it was such a tight fit. I couldn't breathe and it felt like it would break my arms off! During my actual spacewalks both my crewmate and I wanted an XL, but alas there was only one XL on board the ISS (storage is incredibly tight and we only had one M, two Ls, and one XL). I won "rock, paper, scissors," so I went outside in the XL and my crewmate was stuck with the L—there's a great scene in the film *A Beautiful Planet* of Samantha having to pry him out of the suit after our spacewalk. More recently, there was a case of two female astronauts wanting an M, but there was only one on board. Despite the juicy news stories that this made about NASA discriminating against women, the reality is that we just don't have a wide variety of sizes on the shelf because of logistics! Sorry, press.

Next, the limbs are attached to the EMU, including arms, lower body, and helmet. The length of each joint can be adjusted in very fine increments of ⅛ inch. This process takes a few hours, so once a suit is tailored for an astronaut it's a good idea to leave it alone for the remainder of that astronaut's mission. The ability to precisely fit an EMU for each individual spacewalker, making it easier to move around in, is the main advantage of the American suit over the Russian. However, the Russian suit is simple and doesn't require so many hours to set up. Also, cosmonauts can put on their Orlan spacesuits by themselves, without an IV to help them suit up. The downside is that they are bulkier, operate at a higher pressure and are therefore stiffer, and don't fit as well. That all means it takes more effort to move around in the Orlan than in the EMU.

Once the spacesuit is properly sized and organized, it's time to get the equipment ready, an extremely time-consuming process. First, you have to retrieve multiple RETs, AETs, and ERCMs (NASA acronyms for equipment and safety tethers), and they all have to be organized by serial number. Then the actual equipment has to be gathered and placed into crew lock bags or ORU (NASA acronym for piece of equipment) bags, also organized by serial number, and always arranged in a very specific configuration. This process took three of us an entire day for my third spacewalk. Preparing this gear demands total attention to detail; you don't want to end up outside with equipment improperly arranged, or worse, the wrong tool or gear. It's better to spend an hour of time organizing while still inside the spaceship if it saves a minute of time while outside the spaceship.

The actual spacewalk day begins quite early and has a rigid schedule that is similar to launch day. First, you need to go to the bathroom. Because once you get in the spacesuit, all you'll have is your diaper, which is fine for number one, but you don't want to be in a vacuum-sealed spacesuit having just gone number two in your Pampers with several hours to go before you can take it off. You especially don't want to be the IV, pulling a person who just pooped his diaper out of the suit! Therefore, it's best to visit the bathroom before getting suited up.

I had organized my personal items the night prior, so I had a Ziploc bag with ibuprofen that I took prophylactically. It was to prevent at least some of the pain that eight hours in an EMU would cause. Just like in the NBL, I downed a five-hour energy drink and Power Bar and then put on the diaper, thin base layer, and LCVG. Next came an oxygen mask to start purging our bodies of nitrogen. In the same way that divers can suffer the bends, astronauts risk having nitrogen gas come out of their body tissue as pressure drops, which can lead to a multitude of serious medical problems. That risk wasn't a problem with the pressure change we experienced in training underwater, but it definitely was for actual spacewalks. So we went through a long process called ISLE (acronym for breathing 100 percent O_2 while doing light exercise in the spacesuit) to help rid ourselves of N_2. The "exercise" was basically twitching a few times a minute, but it did the job. The combination of breathing pure oxygen and moving limbs around helps to force the nitrogen gas out of our blood and tissues, so that when the spacesuit pressure drops to about a third of sea level pressure, nitrogen bubbles don't work themselves out of our tissue and into painful places like joints, or dangerous places like our brains or lungs or hearts. Knock on wood, the various prebreathe protocols that astronauts have used over the years have all worked. It's a fair amount of overhead in terms of time required before a spacewalk, either breathing pure O_2 from a mask, as in my case, or spending the night at a reduced pressure with the airlock sealed off from the rest of the station, a technique used in prior years, but it's worth it.

With gas mask on, the suit-up began with the help of the IV. First came the LTA (acronym for bottom half of the suit). I found a spot on the floor of the overcrowded airlock to awkwardly hold myself down while pulling on those million-dollar, hundred-pound pants. And to paraphrase Christopher Walken in the "More Cowbell" skit from *Saturday Night Live*, astronauts put their pants on one leg at a time. Except after we do, we go out into space. Getting those things on is a struggle because of the several layers of bulky long underwear, as well as the sticky rubber lining in the LTA. I used small detachable metal handles to gain leverage to pull the pants on. The boots were attached to the pants and were also difficult to get in. I had to point my foot down, like a ballet dancer, and wiggle them down past the rubber bladder. Once both feet and legs were stuffed into the LTA, I stood up and leaned forward, stretching my hamstring like a sprinter getting ready for a race, in order to get my feet deep into the boot.

At this point Samantha, our IV, did the boot/bladder manipulation, which meant she had to unhook the boot from the pants, rearrange the rubber bladder in my boot, and then reattach the boots to the suit. This helped prevent the bladder from being bunched up and causing serious pain in one spot on my feet. An inflated air bladder can feel as hard as steel, and having a pressure point on your foot is a sure way to be miserable for hours. I always appreciated my IV's help with the boot/bladder manipulation because that laborious process saved me a lot of pain!

Next came the HUT. I positioned myself below it, stuck my arms straight up in the air in a surrender pose, and then began a wiggle-worm dance of pushing my body into the suit, right arm, torso, left arm, torso, right arm, head, torso, left arm, inch by inch. The EMU isn't really designed for humans, as the joke goes, but it wouldn't be funny if it weren't partially true. You really have to be a contortionist to fit comfortably into one of those things; my upper body is big and stocky and it just doesn't fit in there well. There is a new procedure that allows the IV to remove the suit arms, which makes getting your upper body into the EMU dramatically easier, but it also takes time to remove and reinstall the arms, so I just dealt with the pain and arm bruises.

Finally, the dreaded helmet. You see, I have a big head. I know— I'm a fighter pilot and astronaut so of course I have a big head. But I mean I actually have a large cranium.

After I squeezed into the HUT, it was time to connect the bottom half to the top half, or the LTA to the HUT. This was usually a two-person job. Astronauts want to be wedged into the suit as tightly as possible, which makes moving around outside much easier, but in order to do that you need a lot of force to squeeze your pants up to attach them to the upper body section. It was often comical to see how hard the IV had to push to get the top and bottom connected. Because I filled out the suit it was a wrestling match to get my HUT and LTA connected, and I really appreciated Samantha's and Anton's work smashing me into the EMU.

Next came gloves. Gloves are the most important appendage to ensure proper fit. Your fingers are constantly squeezing and moving for eight to nine hours during a spacewalk in order to move yourself around, hold equipment, and manipulate tools. I wore a specially cut series of moleskin patches on my fingers and hands to prevent sores and hot spots. Above them were thin glove liners and above them were the actual spacesuit gloves. This required pushing my hand as hard as I could while Samantha held the glove still with all her might, and once my fingers were all the way in she connected the metal glove ring with a distinctive *click* to the suit arm ring.

Each spacewalker has two pairs of gloves in orbit—a prime as well as a backup set, in part because the prime gloves may end up not fitting if your fingers swell, but also because each EVA really takes a toll on gloves. Many hours of holding metal bars and equipment, constant flexing, and extreme temperature swings can chew up the hardened rubber surfaces on the outside of the gloves (called RTV). That rubber on the palm side of your fingers is the part of the glove that gives you grip. After a few spacewalks, the RTV is usually so worn out that the crewmember needs new gloves. What's more, the outside of the station is full of little craters, created by years of being bombarded by small things zooming around space and impacting the ISS. If one of these sharp edges tears the glove RTV, you can keep going—to a point. But if it tears the actual glove fabric, called Vectran, then the glove is no-go and needs to be replaced by a spare on the next spacewalk.

I was always amazed at how chewed up my gloves got after an EVA. I used the same pair for all three spacewalks, but the RTV grip pads were pretty cut up, and if I had done a fourth EVA I would have switched to my spare gloves.

Finally, the dreaded helmet. You see, I have a big head. I know—I'm a fighter pilot and astronaut so of course I have a big head. But I mean I actually have a large cranium. The Russians have measured every astronaut and cosmonaut for their custom-fitted Sokol spacesuits for more than fifty years, and they told me that I had the largest head they had ever measured, dating back to Yuri Gagarin. My F-16 and T-38 jet helmets had to have all of the lining layers removed in order to fit my giant cranium. When putting the Sokol suit on for training, it would always squeeze my face so hard that I would bruise my chin. It was embarrassing—if you Google pictures of me after landing, you can see bruises on my chin.

All of this meant putting that helmet on was a challenge for me. On one particularly difficult day at the NBL, we almost canceled the training run because after ten minutes of trying to smash it on it wasn't budging, until we finally got it on. After years of difficulty, a suit tech finally showed me a handy technique: He put the helmet on while it was rotated 90 degrees, so I was staring at the ear part of the helmet, slid it over my head almost the whole way, and then rotated it back to the normal position before clicking it to the spacesuit. Thank God for that technique; it saved a lot of pain.

Once completely sealed in our spacesuits, we performed a detailed series of checkouts, making sure the pressure controls and communications and cooling and air conditioner systems worked. All while breathing 100 percent O_2 and lightly exercising (twitching periodically).

Our bodies were finally adapted for the imminent reduced pressure we would face outside, our equipment was organized, and spacesuits were ready to protect us from the unforgiving, hellish combination of vacuum, blazing sunlight, and frigid chill of the blackness of space. Getting suited up for my spacewalks was truly a team effort, beginning with our IV Samantha Cristoforetti, but extending to the tremendous support crew we had back in Houston, led by Alex Kanelakos and Faruq Sabur. And most of all, thanks to the suit technician who taught me how to put my helmet on without squashing my brain. I owe you one.

BRIEF THE FLIGHT AND
FLY THE BRIEF

Don't Fly by the Seat of Your Pants

The final days of February 2015 were particularly memorable for me—my three spacewalks happened that week. The first two were largely spent running power and data cables and also putting grease on bolts, as I was Terry the Cable Guy and Grease Monkey. Our third EVA was a completely different ball game, going out to the right and left sides of the station to install reflectors and antennae, as well as cables. To do this we needed an entirely new set of procedures and equipment, and the days leading up to that final spacewalk were some of the busiest of my life.

The frenetic pace led me to have a conversation with our lead flight director down in mission control, the boss for that spacewalk. We agreed not to do any get-aheads, or work that you prepare for in case you finish your normally planned tasks early. Even though they are not mandatory, they require studying and preparing equipment. Because this was our third spacewalk that week, I did not want to add to the list of tasks to prepare for our final trip outside. So "flight" and I came to a mutual agreement—no get-aheads on this EVA. If we finished an hour or two early, we would declare victory, come inside, and high-five each other. We wouldn't add to our workload and prepare for any additional tasks.

Fast-forward to the end of that third spacewalk. Things were going great. We were an hour and a half ahead of the timeline. My spacewalking partner had finished his tasks and was waiting inside the airlock for me to come in and close the hatch. The finish line was in sight when the call came from Houston: "Terry, we have a get-ahead task for you." I had several thoughts—first was, "What the heck, hadn't we agreed on no get-aheads?"

That was really strange. On the other hand, I was feeling good and wasn't in a hurry to go inside. Houston relayed that my task would be to go retrieve an equipment bag that had been left outside years before. How hard could that be? So I agreed to do it.

I should have known better, but because I was feeling good and it wasn't a difficult task I went ahead and did it.

It turned out to be more difficult than I thought, and after about twenty minutes of minor struggle, I had the bag and was back at the airlock, ready to come inside. And whereas twenty minutes prior I had been feeling great, by this point I was tired. Also, because this was my first time as EV1 in the airlock, I would be closing the hatch for the first time, which turned out to be a much more tiring task than I had expected. Doing that task at the behest of Houston caused me to spend precious minutes at risk while solo in the vacuum of space and also dipped into my already depleted reservoir of strength. Not good.

Which brings us to the lesson of this situation. Brief the flight and fly the brief. This was something that I had learned as a fighter pilot and was pretty much unwritten gospel at NASA and every other legitimate flying organization in the world. If you're going to do a dangerous and complicated task, you plan it, brief it, and execute it. If something extraordinary happens you flex as required, but in general you don't fly by the seat of your pants. That lesson has been learned time and again from lethal consequences. In this particular case, I had briefed the flight director that I would not do any get-aheads, but when Houston asked me to do one, I did it. That was bad on me. I was the flight lead, or EV1, the lead spacewalker. I should have known better, but because I was feeling good and it wasn't a difficult task I went ahead and did it. And ended up more tired in the airlock than I should have been. The reality of a spacewalk is that you have to close that hatch manually. And I had just a little less gas in the tank than I would have had if I not done that get-ahead. I should not have agreed to do it when Houston asked.

Fast-forward a few months. I'm back on Earth and doing my spacewalk debrief with the whole team. It was going well as we talked about how we got every task completed on those three EVAs. Then I remembered and asked the question, "Oh yeah, about that get-ahead, what was up with that? I thought we weren't doing those?"

Crickets. Nervous glances around the room, like the kids in *A Christmas Story* when the teacher was asking who made Flick stick his tongue to the frozen pole. So I walked over, shut the door, and then said, "OK, what really happened?" And I heard an unbelievable tale from several of the folks who had been in mission control that day. They described a conflict between the flight control team and the management person in the room, a representative of the FOD (Flight Operations Directorate), ostensibly there to provide guidance to the flight director if the situation called for it. That notwithstanding, the flight director is the boss, in charge of mission control and responsible for the safe execution of the mission.

That day, March 1, 2015, FOD decided to tell the flight director to have the crew perform a get-ahead and go retrieve a storage bag. Flight informed him that we had agreed not to do any get-aheads, and what ensued was a full-blown conflict in mission control, while I was outside, oblivious to what was going down in Houston. When I heard this story, I was furious at the lack of professionalism on the part of FOD, which frankly put the safety of that EVA (i.e., my own pink butt) at risk, but there were no consequences for those who improperly used their management authority in the middle of critical, dynamic flight operations. At the end of the day, we finished the spacewalk well ahead of time with all tasks accomplished.

There is a big lesson to learn from this experience. Brief the flight. Then fly the brief. And let the flight director direct the flight, not management. A lesson that applies to just about any industry.

ALONE IN THE VACUUM

The Spacewalk Itself

The best spacewalking advice I ever got was from a colleague of mine, Rick Mastracchio. He took me aside about four months before launch, while working out at the astronaut gym, and offered me this gem, from an experienced spacewalker to a rookie: "If you're moving slow, you're moving too fast." Simple, witty, grammatically suspect, yet to the point.

It captures one of the many nuggets of wisdom that every astronaut who goes outside must learn. Personally, I have always preferred to learn from others—let them make mistakes and learn the hard way, rather than me! This chapter is a hodgepodge of those lessons, things that you will need to know should you ever find yourself in a spacesuit attached to the ISS with a 140-foot tether pack, hatch open, hearing the words "go for EVA" from Houston.

Let's start with going outside for the first time. I had been in the airlock for close to an hour as we went through the depressurization procedure, we opened the hatch, and my crewmate went outside as I handed him his gear. Even though I was exposed to vacuum, it didn't feel like it. Until I shimmied my body out, feet-first, inch by inch, barely fitting through the hatch diameter. And there I was, witnessing the most gorgeous, spectacular sunrise on the horizon a thousand miles away, a view that you honestly can't imagine. It was still nighttime on the planet below, the blackest black I had ever seen. The first thing I wanted to do was verify that I wouldn't have any vertigo, something that a few of my colleagues had reported. So I attached a tether to the handrail in front of me and let go. Hands free. Floating, watching the Earth floating by at 5 miles per second, 250 miles below, in the early morning twilight. It's a moment that I'll never forget, something that I'm sure is etched in the soul of every spacewalker. And as is the case for so many moments in space, it was immediately time to get back to work.

Before that sublime moment, we had to get all of our equipment ready, so let's start with tethers. Everything that you go outside with needs to be attached to something that is attached to something else, etc. When I watch the GoPro video I shot during my spacewalks, it is amazing just how cluttered the area immediately in front of my chest was. First is the bulk of the space-suit. Then the bulk of the various tools that I carried outside. Then the large, larger, and largest equipment bags that were attached to me or carried by me. Don't forget the camera that began life as an oversize professional camera and then grew to twice that dimension when contained within its special EVA protective covering. The list goes on. There is a lot of stuff you have to keep track of, each item connected by a thin tether that seems to be constantly trying to wrap itself around something else.

There are several different types of tethers. Astronauts themselves use an ERCM (NASA acronym for safety tether), a large, lunchbox-size and peanut-shaped device that contains either 85 feet or 55 feet of braided metal cable. This cable is attached on one end to the spacesuit and on the other end to the ISS, and it is *never* removed from either end without first connecting a new tether. This make-before-break principle is the foundation of spacewalk tether management. A few years prior to my spacewalks, our EVA training team came up with a genius idea—as they are wont to do—to make a safety tether pack comprised of two safety tethers connected to each other in serial, giving the astronaut 170 feet of line. This was a huge improvement over the old system of just using one ERCM at a time, because we couldn't go very far from the airlock before having to lay down a new tether, a time-consuming and painful process. The new safety tether pack allowed us to go from the airlock, the departure point for every spacewalk, to just about any point on the station. Genius.

Having a safety tether is one thing, but keeping track of it is an entirely different ball game. As we crawl around on the surface of the station, hand over hand, we have to route the tether, shepherding it with our hands, wrap-ping it around handrails or other stationary parts to keep it tucked away, close to the station surface. We call that a poor man's fairlead, when you just run the cable behind a piece of equipment, like hanging a string of Christmas lights on an existing post. However, sometimes a more substantial way to keep the tether in place is called for, so we attach a smaller cloth tether to structure and wrap it around the safety tether, keeping it securely in place. That's like nailing in a special hook to hold your Christmas lights securely fixed. The important thing was to know where your tether was at all times and not get it mixed up with your wingman's or tangled with something else. Been there, done that. Even worse would be if your wingman installed a piece of equipment on top of your tether, in which case you'd be stuck outside until you removed the offending equipment.

Most smaller pieces of equipment or tools were attached with RETs, which is the NASA acronym for small tether. These devices were about the size of a flip phone and contained a few feet of cloth string. They were used to hold on to just about everything we brought outside, and we normally used between ten and twenty of them per spacewalk, and sometimes more. RET is not only a noun but also a verb—"RET to that CLB before releasing it from the BRT to transfer to EV2" is a sentence that makes perfect sense to about 400 people at the Johnson Space Center, and precisely nobody else on Earth. It can also be a verb—"the LEE is RET'd to the ball stack." I've never heard it used as an adjective, but it's not inconceivable, especially since I was a math major and wouldn't know the difference.

Another common tether is an AET, which stands for adjustable tether. These are fabric tethers designed to withstand the extreme temperatures and vacuum of space; they're the size of a small belt and can be used to hold things in place. You clip an "adjustable" to structure, pull it tight, and whatever it was attached to is now attached to the station. A large piece of equipment weighing several hundred pounds may be attached to a temporary stowage location using AETs, because there's basically no slack in them. It should also be noted that "adjustable" is a noun only; it sounds very amateurish to use that term as a verb or modifier. For the grammar majors out there . . .

There are a few other exotic tethers, first among them the MWS EE, otherwise known as the grabber daddy, or the mini workstation end effector. It is a big metal claw, the size of a ski glove, on a retractable cloth string directly in front of your chest. Once at your work site you attach it to a handrail, and it keeps you from floating way. I found it to be an extremely useful piece of equipment, though it wasn't a true safety tether. In fact, a friend of mine was using it one day and the clamp mechanism broke apart. If that had been his only method of tethering himself to structure, he would have floated away. Luckily it wasn't. I always used it as a secondary method of tethering, mostly as a way to keep myself in place, like having a piece of string tied from your chest to a structure in front of you.

A similar device isn't really a tether at all, though its official NASA acronym is BRT (body restraint tether). It is a large, 2-foot-long metal hose attached to the left side of the spacesuit. The end is a large, obnoxious metal clamp that requires a pretty good squeeze with your gloved hands to open and is designed to clamp onto a handrail. Once the end is clamped in place and the BRT is oriented in the desired direction, you rotate the end of it, like tightening a hose onto a faucet, and the whole mechanism stiffens up. It's an ingenious system that must have been a mechanical engineer's dream to design. Though it takes a bit of muscle to manipulate, once it is tightly clamped and stiffened you are held in place fairly well, allowing you to use both hands for work. It's not as strong as having your feet in an APFR (NASA acronym for foot restraint), but it's very handy and only takes about thirty seconds to get it set up, whereas the APFR might take thirty minutes or more to retrieve, set up, and stow when you're finished.

Besides managing tethers and equipment, you have to be able to find your way around while outside. This may not seem like a big deal, and when the sun is out, it's a fairly simple task to look around and see where you are. However, at night it's an entirely different task. It is dark outside in space at night. I mean *dark*. And though your headlamp is on, it's a narrow-beam flashlight that shows only what's directly in front of you, like the spacesuits in most space movies. There are usually a few station floodlights on at various locations, but they aren't intuitive and definitely don't give you the big picture of which way is up or down. It's like having a porch floodlight on in a very big backyard on a very dark night. So you really need to keep track of where you are. We call this having SA, or situational awareness. It's a skill useful in all aspects of being an astronaut, or fighter pilot, or life in general. There are

Only a millimeter of plastic between my face and the deadly vacuum of space.

some pretty funny stories of guys getting lost. They basically had to crawl around until they came to something they recognized, which can be embarrassing and time-consuming. Thankfully, it never happened to me, but who am I to judge? There are those who have and those who will.

It's also interesting to go through a day/night cycle while outside. Mission control gives spacewalkers a heads-up that the sun is about to come up or go down. If you ever listen to the audio of a NASA spacewalk, you will hear the occasional, "Two minutes to sunrise/sunset." That lets the crew know that they will have to raise or lower their sun visor shortly. There is also a set of lights on the helmet that you could theoretically turn off and on every forty-five minutes, but I just left mine on the whole time. Raising my hand up to the top of my helmet and pushing buttons to turn on lights required time and muscle energy that I wasn't willing to expend, so on they stayed.

When the sun is up, it is *hot*, +250 degrees they say. I assumed Fahrenheit, though nobody ever specified. And when the sun was down, it was *cold*, –250, which must have been Fahrenheit because –250°C would be basically absolute zero. One of the impacts of the cold night is that your fingers could get very cold, like when skiing, so NASA added some battery-powered glove heaters. They turn on with a cloth tab on the back of your gloves, but this required a fair amount of effort to grab and yank. Plus my fingers never really got cold, so I never bothered with the heaters. Except one time.

Before each of my spacewalks, we had a briefing from our ground engineers about when they predicted that we would be cold or hot. For nearly all of my three spacewalks, more than nineteen-plus hours outside, I was predicted to be comfortable, with the exception of one cold and one hot moment. I distinctly remember the first moment; I suddenly got very cold and recalled that my EVA engineers had predicted that precise time and location. Impressive! I was about to reach for the glove heaters when I noticed the sun about to rise, so I waited a few seconds. Sure enough, as soon as that blinding light popped above the horizon my chills instantly vanished. It was impressive. Millions of tons per second of nuclear fusion does a pretty good job of keeping the solar system warm! It made me realize just how cold most of the universe must be, far from the warmth of a sun.

> I felt like an ant getting melted by a twelve-year-old with a magnifying glass, only in this case it was God melting me.

On a different occasion, I was on the very front of the station, at the forward end of Node 2; it's where PMA-2, the module where my space shuttle had docked, is attached, and where future human capsules will dock. Suddenly, I felt a heat unlike anything I had ever felt; it was like pins and needles. My body intuitively knew it was heat, though it was a different sensation, like infrared energy as opposed to a normal sensation from hot air. It was at the exact time and location the engineers had predicted. I moved a few feet, away from the infrared energy that was pouring off the jet-black surface of PMA-2 and being reflected by the shiny aluminum of Node-2, and was immediately comfortable again. It was fascinating to viscerally feel this extreme thermal environment in my body, if only for a few moments. I felt like an ant getting melted by a twelve-year-old with a magnifying glass, only in this case it was God melting me.

There were two basic tasks that I needed to get done on my three spacewalks. First was laying cable, just like Larry the Cable Guy. Our job was to route power and data cables from the station's central hub out to the docking ports and radio antennae that future capsules will use. We eventually installed more than 400 feet of cables, the most ever done in one project. This low-tech task involved bringing out large bags full of cables that were prelabeled and stored in the precise order and orientation in which they would need to be removed, and then attaching the bag near the start point. The first step was to plug one end of the cable to its electronic box. After that the cable was secure—no need to tether it, alleviating a huge pain. Next, we rolled it out of the storage bag, fixing it to the station every 10 feet or so using wire ties, roughly 2-foot-long pieces of stiff wire that wrap around fixed structures like a twist tie on a loaf of bread. Except they work in the extreme environment of space and cost a few hundred dollars each. A low-tech solution to a high-tech problem—the epitome of innovation.

My other task was lubricating the station's robotic arm, which had been outside in the extreme space environment for over a decade and had some joints that were getting creaky. This task involved close coordination between myself and Samantha Cristoforetti, who was inside flying the arm. I was sta-bilized in an APFR next to the airlock, grease gun and lube tool ready to go, while Samantha maneuvered the arm right in front of me. I would put grease on the arm's sticky bolt or mechanism, and then she would quickly rotate the joint to the next position. Because she was so fast and efficient at flying the arm, we were able to get all of its key parts lubricated in our allotted two and a half hours. I was told that this was the first time a crew had gotten through each of the required tasks, and it was all because of Samantha.

Applying the grease was a fairly straightforward process that was not without a few laughs. Let's begin with the lube tool itself. Remember the wire ties? Well, we took a wire tie, straightened it out, duct-taped a screwdriver to one end to form a handle, and bent the other end into a V-shaped tray, which was completed with, you guessed it, duct tape. This became our lube tool, another low-tech solution to another high-tech problem. With a giant grease gun in one hand and the lube tool in the other, I squirted grease into the V-shaped tray for the first time and looked away to holster the grease gun. When I looked back up, the grease was gone from the lube-tool tray. I looked around to see if one of my crewmates was there playing a joke on me, but of course I was alone. I didn't think much of it, put some more grease in the tray, and continued with the procedure, deliberately lubricating each of the sticky bolts in turn. An hour and a half into this process, I was once again holstering the grease gun, only this time in the corner of my eye I caught the grease slowly floating out of its tray, into the darkness of space. I think I yelled "Nooooooooo" in slow motion, immediately realizing what had happened to that first wad of grease. I told Houston, and they told me not to worry about it, just put some more grease in the tray and press on.

Fast-forward a few months later, after I was back on Earth. I was sitting at my desk answering emails when a colleague who worked in the station program walked in and said, "Terry, you might find this interesting." He had a photo of the exterior of the station, which the new crew in space had taken, showing one of the starboard radiators. Its series of big, white, flat panels had a blemish on it. As we zoomed into the JPEG it was clear; that brown spot was a blob of grease. We laughed so hard—I had left a permanent mark on the exterior of the station, and it was awesome.

I used to say that of the 500 individual tasks that had to be accomplished on every spacewalk, 499 were optional. The only must-do was closing the hatch after your spacewalk. Even if you couldn't open the hatch at the beginning of a spacewalk, that wasn't the worst thing that could happen. Although you might be mad, you would be safe inside the ISS. But if you couldn't close the hatch at the end of a spacewalk, you couldn't repressurize the airlock with air, and you'd be stuck outside with the rest of your life to figure out how to get that hatch closed. It also turns out that the hatch can be particularly finicky; I've since heard tales from other crews who almost wore themselves out trying to get the hatch open or closed. Unfortunately, I didn't hear those stories until after my flight! At the end of my third spacewalk was the first time I had to close it, and it was a real struggle. I was getting winded.

To visualize the task of closing a hatch, imagine yourself lying on your stomach on the floor, looking down, and in front of you is a big, window-size hatch. You need one hand to hold your body in position because you're floating, one hand to hold the top handle on the hatch, and another hand to hold the bottom handle, because the whole hatch is basically free-floating and needs to be stabilized. You also need another hand to rotate the latch to seal the hatch once it's closed. It was a struggle. I would get the top of the hatch pressed against the seal, and then the bottom would pop up. I would grab the bottom handle and push it down, but then the top would pop up. A third hand sure would have been nice. This whole time my other crewmate was completely useless, because he was wedged into the other side of the airlock, head in the other direction, unable to see or reach anything I was doing. Eventually, I figured out how to push the whole hatch firmly against the seal; next, I had to grab the latch and rotate it to the locked position. By this point a fourth hand sure would have been nice. The latch had an arrow indicating clockwise, so I rotated it clockwise and could feel the mechanism moving, but not catching. I kept on rotating it, struggling to keep the hatch firmly in place against the seals, rotating the knob, sweating, and burning up a lot of energy, fighting against the pressurized and bulky spacesuit.

Finally, I read the label on the arrow. Clockwise was the direction to open the hatch, not close it! I thought (not out loud) some bad words, rotated it counterclockwise, and voilà, the hatch grabbed and firmly sealed. And I wondered—why in the world, if you had only one label for one arrow direction, would you make it in the open direction? I laughed quietly to myself, we repressurized the airlock, and all was well. But I learned a few lessons. Build hatches that don't require four hands to close and have arrows that point in a smart direction.

During airlock repressurization on my second spacewalk, I noticed water pooling on my helmet visor, which seemed normal, probably just sweat dripping off my face and falling onto the visor, since I was looking down. Then I remembered—I was in space, and water doesn't fall onto anything; it floats! The pool of water was growing bigger by the minute. The spacesuit I was in, serial number 3005, was prone to having a small amount of water leak out, especially during repress. But this blob was bigger than just a few drops and it was still growing. This was a very high-emphasis item at NASA, because two years prior one of my colleagues, Luca Parmitano, had nearly drowned in his suit during a spacewalk. After that close call there were a lot of procedures put in place to mitigate that risk, including installing a snorkel inside the suit and adding an absorbent pad in the helmet.

The water grew until it covered my entire visor and the back of my head was squishy wet, and I finally made the call. "Houston, EV2, I've got some water in my helmet. It is probably related to the issues that 3005 has, but I wanted to let you know because it's covering the visor and I can feel it in the back of the helmet." I was hesitant to make that call because I knew it would get folks on the ground spun up, including the press, and I was right. While my daughter was driving home from school that day, she heard on the radio, "Astronaut Terry Virts is on a spacewalk and has water in his helmet and might drown." Exactly what I didn't want to happen. But I was proud of how calmly the flight control team in Houston handled it. We proceeded with normal airlock repressurization and I got out of my suit quickly, without rushing too much.

> The only sound I heard was the faint, high-pitched whine of the spacesuit fan, and my own breathing, and for a few glorious seconds it was just me and the universe.

Our crewmate Anton Shkaplerov floated down from the Russian segment and helped Samantha expedite things. He also used a syringe to measure the exact amount of water that had pooled in my helmet. That measurement allowed me to go outside in that same spacesuit, number 3005, three days later because our engineers were able to narrow down the source of the water to a benign type of leak. Had we not had that syringe data I think NASA managers would still be having meetings to this day, trying to decide whether or not to send me out on the spacewalk.

Ninety-nine percent of my time outside during spacewalks was spent working. I almost always had a face full of equipment and station structure, and I was constantly keeping track of gear, tethers, and the to-do list. I have never felt so on-the-clock as I did during my three spacewalks; there was no time to rest or to pause and take photos. However, during one particular moment on my second EVA, I was at the front of the ISS and had a few seconds to rest. I took that opportunity to rotate my body around and look away from the station and out into space. What I saw changed my perspective on life. There was the most gorgeous sunrise, stretching from horizon to horizon and filling my field of view, beginning as an intense blue to the right and morphing into distinct lines of orange and red and pink. Below was the Earth, black as coal. Above was infinity, blacker than the darkest night you've ever seen. The only sound I heard was the faint, high-pitched whine of the spacesuit fan, and my own breathing, and for a few glorious seconds it was just me and the universe. I felt like I was seeing God's view of creation, something that humans were not meant to see, and I could hear Him tell me, "I am." That's all, just "I am." Adjectives have not been invented to adequately describe this moment, so I won't torture our language by trying, but you can do your best to imagine.

And then I had to get back to work; there was a power cable that needed to be connected to a cable tray on PMA-2 that would eventually be connected to the capsule docking ring.

You get the point. That moment was a microcosm of my seven months in space. A continuous juxtaposition of the sublime and the mundane, from those first eight and a half minutes during *Endeavour*'s launch to the end of my 200-day mission, 99 percent of my time was spent repairing equipment and storing gear and putting grease on bolts and running on a treadmill. And 1 percent of it was spent hearing from God and seeing creation from a perspective that I'd never thought possible.

So if you're planning a spacewalk, remember these things: Keep track of your tethers. Don't bobble the grease tool. Rotate the hatch knob counterclockwise to shut it. Take a few minutes to look out into the universe and hear from God. Water doesn't fall down in space. And above all—if you're going slow, you're going too fast.

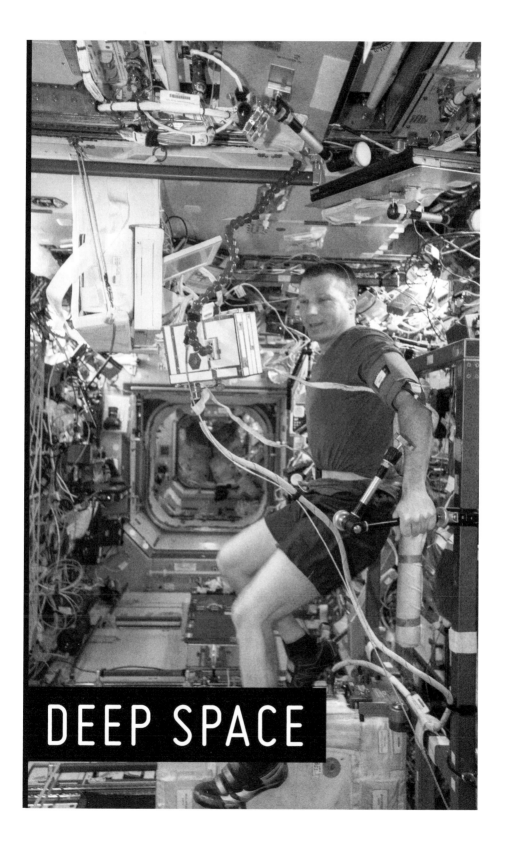

DEEP SPACE

WHAT YOU NEED TO GET TO MARS

A Realistic Look at What It Will Take

orty years of multinational, long-duration missions on the ISS, Mir, Skylab, and Salyut have really demonstrated that people can live and work in space for long periods of time. Russian cosmonaut Valery Polyakov even flew a mission that lasted more than 437 days! Proving that humans can thrive during missions lasting half a year or longer has paved the way for future human exploration of our solar system. However, those missions will require long transit times to their planetary destinations, during which the astronauts will be exposed to radiation and weightlessness. Even though we have conquered the standard six-month low-Earth-orbit flight, there are lots of challenges that need to be met if we are to leave the relative safety of Earth and venture out into deep space.

The twenty-first-century goal for human exploration is to get to Mars. The Moon will be an important proving ground to test out the equipment and technologies we will need to get to Mars, but the long-term goal will be the red planet. It is a much more interesting and hospitable destination than the Moon in several key ways. A day on Mars, from sunrise to sunrise, lasts twenty-four and a half hours, very similar to our home planet. A day on the Moon, however, lasts more than twenty-nine Earth days. There is an atmosphere on Mars, albeit very thin, that could be useful. There is water, frozen in the polar ice caps. There may have been oceans on Mars in the past. Mars is much more likely than the Moon to potentially harbor life, if only microbial. The gravity is twice as strong on Mars as on the Moon, and much closer to Earth's gravity. The radiation environment is much better on Mars because it is farther from the sun. The soil on Mars is similar to that of an Earth desert and could be used for farming, while the soil on the Moon is extremely harsh, more like crushed glass. And the list goes on. If there is an interesting

Thankfully, electric propulsion engines can fly much faster than chemical ones and would enable a one-year mission to Mars: four to six months outbound, a month or two on the surface, and another four to six months for the return trip.

destination in our solar system for humans to visit in the twenty-first century, it's Mars.

Now that you're convinced that we should send astronauts to Mars, what technologies do we need to develop to go there? Why don't we just fly there? There are plenty of Hollywood movies where the big spaceship magically appears and off the astronauts go to the red planet, most recently *The Martian*. Let's dig into the details of what we need to make this journey happen. The list is significant.

Many types of equipment need to be developed: landers, rovers to carry astronauts, water and air recycling systems that are more efficient and reliable, spacesuits that can be used multiple times in a dusty environment and easily maintained by the crew, lightweight exercise machines, helpful robots, light bulbs that don't burn out, cleaning equipment that isn't massive, etc. Equipment reliability is a big issue; twice during my 200-day flight I spent an entire week repairing the carbon dioxide removal equipment, using many bulky and heavy spare parts. Those critical systems need to get more reliable and lighter, and maintenance is an area in which perhaps 3D printing can help. These aren't insurmountable issues, and our efforts on the ISS and eventually the Moon should be focused on improving equipment in order to enable missions to Mars.

Beyond all of these, there is one overarching piece of technology that needs to be developed—nuclear power. This will serve two purposes: enabling electric in-space propulsion and providing crews with electricity while on the planetary surface. NASA probes have been using nuclear power since the 1960s, generated from RTGs (radioisotope thermoelectric generators). They use a few kilograms of plutonium or other radioactive material that heats up while emitting low-level radiation, warming thermocouples, which then convert heat into electricity. A typical space probe RTG generates a few hundred watts. Because plutonium has a half-life of more than eighty-seven years, they last a long time. In fact, the Voyager probes launched in 1976 still generate roughly 200 watts of power from their RTGs, enough to operate some basic instruments and send very weak radio signals to Earth, despite having actually left the solar system.

These devices are simple, safe, and very reliable. However, they generate only hundreds of watts and are therefore useless for human exploration. We need megawatts to drive in-space engines and kilowatts to run surface life-support systems.

First let's discuss propulsion. Conventional chemical engines, the kind that rockets have been using for decades, burn a fuel and an oxidizer in a chemical reaction. This type of traditional propulsion requires three years for a human mission to Mars: six to nine months outbound, a year and a half on the surface while waiting for Earth and Mars to orbit the sun and align again, and then another six to nine months for the return trip. I believe that three years is too long; it's too many supplies to pack, too much radiation exposure for the crew, and too much mechanical malfunction risk. Thankfully, electric propulsion engines can fly much faster than chemical ones and would enable a one-year mission to Mars: four to six months outbound, a month or two on the surface, and another four to six months for the return trip.

Electric rocket engines are nothing new. They have been flying in space for decades, though on a much smaller scale than that required for a human mission. Instead of burning a fuel and oxidizer to shoot exhaust out of the nozzle like a conventional rocket does, electric propulsion uses an electric field to accelerate a charged propellant (ionized gases such as hydrogen or xenon or others) out the nozzle, which may be a magnetic field to contain the ionized gas instead of a traditional metal nozzle. Using this type of engine takes advantage of something called the rocket equation, developed for space travel back in 1903 by the Russian scientist Konstantin Tsiolkovsky. This equation states that the speed at which a rocket can travel is directly related to the speed of its exhaust velocity. Because ionized gas shoots out of the nozzle tens of times faster than traditional rocket exhaust, an electrically powered spaceship can theoretically travel much faster. But there is a rub. . . .

It comes back to that question of electrical power. In order to propel a relatively small satellite, you can use solar power to generate your electricity and use a small electric engine. In fact, this propulsion technique has been used on more than 200 satellites over the past several decades. But people require big, massive spaceships. And in order to make a useful electrical engine that could really get the behemoth moving at the speeds required to shorten that Mars trip to one year, we would need a nuclear reactor generating around 50 megawatts of power. Although reactors of that size are fairly common on Earth, they've never been made to that scale for space. Until we have the political will to build those nuclear power plants in space, we will be limited to a three-year round trip to Mars. Which has a tremendous number of implications, none of them good.

First, supplies. You are going to need a lot of food and underwear and water and toilet spare parts if you are going to be gone for three years. You need a lot for a year also, but three-year round trips would require, well, three times as much. Except the dollar cost is more than three times greater because you need more supply rockets to launch from Earth. You need a much bigger transfer vehicle to take you to and back from Mars. You need a significantly bigger surface presence, with multiple additional modules, to enable a 500-day stay on the surface versus a 50-day stay for the fast trip. Every additional pound of surface equipment and supplies will increase the demand for very expensive landers, which also must be launched from Earth on a greater number of very expensive rockets. Et cetera. Adding two years to a mission is not a linear increase in cost; it is exponential.

One of the most accurate quotes from *The Right Stuff* is, "No bucks, no Buck Rogers." The early Mercury program astronauts had grasped a fundamental truth of spaceflight, perhaps more important that Mr. Tsiolkovsky's rocket equation. The real fuel that makes these spectacular missions fly is money. So let's talk about the cost of a three-year mission using traditional rockets. NASA's preferred heavy-lift rocket is known by its very sexy acronym, SLS. Space Launch System. Though some prefer *Senate* Launch System (more on that later). We have been spending roughly two billion dollars per year on this rocket, going back to 2005 when it was first conceived as Ares V. When it finally flies, maybe in 2022 or 2023, the cost of each launch will be roughly one billion dollars. Let's forget all of the development costs; they are sunk. I have been told by mission designers at NASA that a single three-year mission to Mars will require seven SLS launches at a minimum. But wait—there's more! The current expected flight rate for SLS is one launch per year. So, let's say the seven-launch estimate for one crewed mission is correct, and let's assume that they can double the flight rate to two per year, which would certainly increase the cost. That means it will take at least three and a half years just to build the Mars ship in Earth orbit before it departs for Mars, with launch costs of seven

Each subsequent crew would leave a module behind, gradually building a human presence on Mars, in much the same way that we slowly built the ISS, piece by piece.

billion dollars minimum, plus the costs of the landers and modules and spacesuits and equipment and mission control personnel and astronaut salaries (wishful thinking). Because we have been running trillion-dollar annual deficits, I'm afraid our debt-strapped government won't have unlimited resources in the future for space exploration. We need to find a faster and less expensive way.

There are several other heavy-lift rockets potentially on the horizon. Elon Musk's SpaceX has flown the Falcon Heavy rocket several times and recently sent its first live crew to the ISS, after it famously launched "Starman" in his Tesla toward Mars. SpaceX is also developing a much bigger rocket that was initially called BFR (Big Falcon Rocket or Big F*!ng Rocket, depending on who you ask) and recently renamed Starship, with a first-stage booster rocket called Super Heavy. Jeff Bezos's Blue Origin is developing New Armstrong, another massive rocket. All of these promise a lift capability approaching that of the SLS for much less than half the cost. We will see if they actually fly; it's very easy to make promises in the rocket business but an entirely different thing to deliver. There are also several other medium-lift options, notably Elon's Musk's Falcon 9 and Jeff Bezos's New Glenn, as well as United Launch Alliance's Vulcan and Northrop Grumman's OmegA. A Mars mission would likely include a series of launches, using different rockets to assemble a Mars transfer vehicle in less than a year. The crew would then launch, rendezvous with it, and head off for the red planet.

Launch concerns aside, I believe two greater problems we need to address are minimizing the crew's exposure to radiation as well as the equipment's exposure to failure, both of which are helped by a fast-boat, one-year mission. The nuclear-powered spaceship would accelerate for the first half of the flight at a very slow but constant rate. Then the whole spaceship would turn around and decelerate for the second half. There would be a heat shield to help slow the craft down when it arrived at Mars, using atmospheric braking, and then the transfer module would remain in orbit while the crew went down to the surface for a month or so. They would descend and land next to a surface module that was waiting for them. Because they are unmanned, those prepositioned supplies could be sent via a slow-boat approach, saving money on launch costs and in-space nuclear propulsion. Each subsequent crew would leave a module behind, gradually building a human presence on Mars, in much the same way that we slowly built the ISS, piece by piece. The crew would return to Earth in the same transfer vehicle that took them to Mars, and they would use the heat shield again when arriving home.

The nuclear power plant, engine, and habitat module could be used again and again, perhaps for decades, cycling back and forth between Earth and Mars. The planets align for launch opportunities every twenty-six months, so a single transfer vehicle could theoretically support four missions per decade, each one building on the last one, until we had an infrastructure of surface logistics modules, an operational nuclear reactor on the surface of Mars, and a reusable and refuelable lander, transferring crews to and from Martian orbit to the surface.

In this manner, crews would be exposed to the harsh radiation environment of space for only one year instead of three, and they would require only one year's worth of supplies, saving a tremendous amount of money and cancer treatment. The technology that enables this plan is nuclear-powered in-space electric propulsion. There are significant technical challenges to fielding such a reactor, but they can be overcome—this is not unobtainium.

In addition to the electrical power problem, there are other systems that must be developed: landers, transfer vehicles, better and more reliable life-support equipment, etc. Those problems are real but are also solvable.

The overarching challenge is finding and maintaining the political will to make in-space nuclear power happen. As I have said many times, the problem isn't the rocket science, it's the political science. I firmly believe that this nuclear option doesn't simply make a Mars mission more attractive, it actually enables such a mission. Conversely, without a way to get there fast I don't believe a three-year mission is possible. It's either electric propulsion or we don't go anywhere in the solar system beyond the Moon. The choice is that stark.

I would choose to go to Mars and other planets, not because it is easy, but because it is hard. (That sure does have a familiar ring to it.)

THE HUMAN BODY BEYOND EARTH

The Physical Toll of Long-Term Spaceflight

One of the most fascinating parts of leaving Earth was observing what it was like to adapt to weightlessness. I was constantly amazed at how quickly and completely humans were able to survive and thrive in that very alien environment. Our bodies were designed, adapted, and evolved to live in Earth's gravity. Then, roughly eight and a half minutes into your rocket ride, the engines shut down, someone turns the gravity switch off, and you are floating, completely removed from the only home humans have ever had. Though the process of adapting to weightlessness is fascinating, it's not all fun and games, and there are some serious medical issues that astronauts face. And some funny ones, too.

The first issue that everyone has to deal with to one extent or another is SAS (Space Adaptation Syndrome), the NASA acronym for feeling sick. Just about every colleague of mine has reported feeling some type of dizziness, headache, or downright nausea, especially on their first flight. For me personally there were two main impacts—a strong dizziness and a headache. I could hardly rotate my head for the first day or two in space; quickly looking left or right or up or down was extremely painful and felt like it would have instantly made me barf.

Speaking of barfing, NASA has some very expensive emesis bags that are amazing. It's a basic plastic bag with a big washcloth sewn into it with an integrated twist tie, so you simply open it up, barf, wipe yourself off, tie it up, throw it away, and get back to work. Some astronauts have reported suddenly throwing up without any warning several days into the mission, so I kept one of these sick-sacks in my pocket for the first week or so, just in case. In fine government procurement tradition, I was told that these cost $500 each, so NASA was a little stingy with them. When I flew on the Soyuz the Russians didn't have anything like ours, so I requested two from NASA, just in case. It was a battle with astronaut office management to get approval, but I eventually won. I thought it was a no-brainer to let our crews fly with a decent barf bag. Better safe than sorry, especially when "sorry" includes making a floating cloud of vomit inside a hundred-billion-dollar laboratory for your five other buddies to avoid.

> It turned out that this pain was from the same phenomenon that was affecting my feet— my nerves had stretched at different rates than the rest of my body, and I felt it.

I also had a serious headache on my first flight, as many astronauts do. I was downing ibuprofen left and right, but it just didn't help. On the morning of flight day three, my crewmate "Stevie-Ray" Robinson asked me, "How are you feeling?" I told him I still had a bad headache, and then a few minutes later it was like someone flipped a switch and I was magically cured. I never had another problem with that headache/dizzy/SAS feeling for the rest of that two-week mission or on my next mission four years later. My brain was very confused for those first two days in space, but then it just magically figured out weightlessness.

SAS ties in with another problem that some astronauts experience: insomnia. Thankfully, there was a solution that helped with both—Phenergan. It's a common medication on Earth, used to prevent nausea for lots of reasons: morning sickness, motion sickness, post-chemo nausea, etc. NASA used a composite medication called Phen/Dex to help during our zero-g aircraft flights; the Phen would prevent nausea and the Dex would keep you awake. Hence, the real benefit of Phenergan for astronauts—without the Dex it helps us sleep, which is especially helpful on that first night or two when anxiety, nausea, and the need for sleep are all high.

There are several ways to take this medication: as a pill, shot, or suppository. The doctors are big fans of the suppository method because apparently things get absorbed into the bloodstream very well down there. As a self-respecting fighter pilot, I turned that technique down. There was also the pill form. Though I never took any, I carried a few Phenergan tablets in a Ziploc in my pocket, so I wouldn't have to scavenge for them. The problem with this was twofold; first, pills are the slowest method to achieve the desired benefits, and second, if you barfed up the pill, well, that wasn't very helpful. Which brings us to the third method—shot. In the butt. Which was what many of my colleagues chose.

On that first night in space, after an incredibly exhausting first day, with the wakeup call only six hours away and sleeping bags velcroed all over *Endeavour*'s walls and ceilings and floors, we lined up for the shot. In training we had learned the technique—aim off to the side of the cheek so as to avoid a long and very painful nerve that goes through the buttocks. Our flight surgeons had marked an "X" on everyone's butt with a Sharpie before launch to help us avoid the nerve. I'm not kidding. So we each duti-fully floated by our Crew Medical Officer, Velcro-laden shorts hovering at the knees, and he jabbed the needle all the way to the base into our butts. Alcohol swab, Band-Aid, "Next!" All was well, and we all went off to sleep without incident. Fast-forward five days. I was rushing from the station back to the shuttle to get a checklist for a maintenance procedure. And my butt itched. I scratched it, but that didn't help. So I reached down into my floating shorts to give my cheek a better scratch—and lo and behold, out came a Hello Kitty Band-Aid. I had forgotten and left it stuck to my butt the whole time! We all got a good laugh about that, and I had a memento of my Phenergan shot.

A very significant effect of weightlessness is bone and muscle degrada-tion. On Earth, everyone's bones and muscles are constantly working against gravity, even when lying in bed or sitting on the couch watching *Game of Thrones*, but in space you never get that passive work. You are floating 24/7, which means that your bones and muscles need help to prevent serious atro-phy. Which is why the various space agencies have developed a very effective exercise and vitamin D protocol over their decades of shared, long-duration space experience. In fact, I came back to Earth after 200 days in space having lost 0.0 percent of my overall bone density, and in good muscular shape. Of all of the medical problems space travelers face, bone and muscle atrophy is not one of them, thanks to the exercise and diet protocols developed and verified on the ISS.

VIIP (NASA acronym for eyesight problems) has received a lot of attention in recent years. A few returning astronauts have experienced degradation of their eyesight. The cause of this has been debated, but a likely contributor to astronaut vision problems is the fluid shift that occurs in weightlessness. A lot of water that is normally in your lower body floats up in space, with much of it ending up in your head. Which leads to a puffy face, making many astronauts nearly unrecognizable compared to their Earthbound selves. It would be interesting to study how that affects facial recognition software. But back to eyes.

A lot of our time in space was spent studying how fluid shift affected our eyesight, with its attendant increase in intraocular pressure (translation—our eyeballs were squashed). Like most astronauts, my short-distance vision worsened and my long-distance vision improved because the shape of my cornea and lens changed. My eye doc even sent up an anticipated prescription set of eyeglasses, assuming my vision would change once in space. He was right, though I never used the prescription glasses. The good news for me, as with the vast majority of astronauts, is that my vision returned to normal when I got back to Earth. The bad news is that for a few, their vision is permanently degraded. Though fluid shift is a prime suspect, other factors such as age, gender, and even elevated atmospheric carbon dioxide levels may be contributors. For now, the studies continue and, knock on wood, astronauts continue to return from spaceflight with functional vision.

Skin problems were some of the most pervasive (and gross) problems we had. Everyone on Earth develops calluses, mostly on feet, but also occasionally on hands, elbows, etc. Where there is a lot of constant pressure, skin hardens and develops calluses. When you get to space most of that contact and pressure stops, so the calluses slough off and disappear. Interestingly, though, while calluses on the bottom of your feet disappear, some astronauts grow them on the tops of their feet, where their feet are often wedged under metal handrails in order to hold their bodies still in weightlessness.

On my tenth day in space one of my crewmates told me, "Hey Terry, float here and tap your foot on the floor." That was a weird thing to do, but I tried it. Wow! Electric shocks shot through my legs into my body. I had never felt nor even heard of such a thing. For the rest of my time in space, I would occasionally tap my feet on a hard surface and . . . *Zap!* An electric, nerve-tingling sensation. A few weeks after my first flight I asked a fellow astronaut who is also a medical doctor about this and he said, "It's rare, but occasionally folks have reported this. In fact, one astronaut returned from a flight a few years ago and still has the problem today—his leg is still numb." How the heck did I spend a decade as an astronaut, talking to nearly everyone about their space experience, being trained as a Crew Medical Officer, working with NASA docs, and not hear about this?

Apparently, this effect was due to nerves and muscles stretching at different rates as your body grew in weightlessness. My height increased by 2 inches without gravity pushing me down; I was finally 6 feet tall! Unfortunately, I was back to 5'10" a few hours after landing, but it was fun while it lasted. Although the added height was nice, the nerve pain was not. One problem that rookies tend to have when adapting to weightlessness is that they grab handrails and pull themselves too hard, and I was no exception. I was having to make a conscious effort to be more delicate as I pushed off and floated from point to point. After a week of learning how to move around, I began to have some serious chest pain—it felt as if I had torn my pectoralis (chest, or moob, muscle). If I pulled on a handrail to launch myself across the module at the wrong angle, there was a sudden and sharp pain, excruciating, if only for a second. It turned out that this pain was from the same phenomenon that was affecting my feet—my nerves had stretched at different rates than the rest of my body, and I felt it.

This same sensation returned on my second, long-duration mission, and it wasn't a lot of fun. On that flight I did quite a bit of weight lifting, and as I'd put my feet on the floor in order to push up during a squat or dead lift, that same electric, burning feeling returned. I was able to work through it, but it was definitely an unexpected physiological effect of spaceflight for me, and it continued for months.

Although this was a nerve issue and not a symptom of a more serious muscle problem, the amount and intensity of our workouts on the station were a cause for concern for me. First of all, you can put a *lot* of force on the

ARED machine, up to 600 pounds, and that's potentially dangerous, especially when both you and the machine are floating. There is a specific technique to operating ARED safely and I took my time building up to higher weights; I didn't want to get smashed unexpectedly, because there's no hospital in space to help with broken bones. It's also possible to overdo it on your muscles and tendons and joints. I was very careful to avoid this, because an injury would mean an extended length of time with no exercise, and I didn't want to miss it. Fortunately, I was able to work out nearly every day of my 200-day mission.

Despite my caution, one day while squatting I felt the dreaded tweak in my back. I stopped exercising immediately, but over the following twenty-four hours the pain continued to grow. I ended up going two weeks on a reduced exercise protocol to let it heal. Overall that wasn't too bad; thankfully, I had stopped exercising as soon as I felt it or it could have taken much longer to heal. But it made me realize that you shouldn't push it in space, because being unable to work out can really impact your overall health. Those exercise machines aren't there just to make you look good at the beach; they're there to combat the continuous, relentless, degenerative effects of zero g.

All of the medical issues I've discussed so far are interesting and annoying but can be dealt with. However, there is one giant, overarching problem that makes flying humans in space a dangerous proposition. Ionizing radiation. Galactic cosmic radiation. Exotic radiation that doesn't exist on Earth. In quantities that don't exist on Earth. When you talk about sending humans out into the solar system (as I believe we should), radiation is the elephant sitting on your couch in the living room. It's a problem that we haven't yet figured out how to solve, nor do we understand its extent.

I experienced radiation in a visceral way, by seeing "white flashes" when my eyes were closed, especially when the station was over the SAA (South Atlantic Anomaly), a weak point in Earth's magnetic field and an area filled with a high level of cosmic radiation. The first time I saw this phenomenon was on my fifth night in space; when I closed my eyes for bed, I saw a brilliant white flash for just an instant, and I thought, "*Cool*—that's what the Apollo guys were talking about!" Then I realized what was happening. If one particle was hitting my optic nerve, that meant there were a thousand other particles hitting other parts of my body, each one potentially damaging the DNA of my cells, which could potentially lead to cancer. Then it didn't seem quite so cool. I eventually saw those white flashes tens, if not hundreds, of times. And every time I saw them and checked where we were over the Earth's surface, we were over the South Atlantic Anomaly. It was an unexpected as well as ominous method of navigation, to be sure.

The really scary effects of this are unseen. Those particles, originating at our sun or other stars in our galaxy or even in other galaxies, some moving at nearly the speed of light, some massive atoms or molecules, sped through the walls of the ISS unimpeded and impacted the cells in my body. Sometimes they would kill a cell, and sometimes they would zip right through my body. But sometimes they would impact the DNA in a cell, altering it at the molecular level. Hence the term *ionizing radiation*. And when that happens, one of the possible outcomes is that the cell would start to mutate, becoming cancerous.

Unfortunately, we have data points from the effects of acute radiation, from the survivors of Hiroshima and Nagasaki. Many of those survivors were stricken with cancer, and one of the most common was basal cell carcinoma, a less-serious type of skin cancer. After both of my spaceflights my dermatologists found this in my skin—quite a bit, unfortunately, after my long-duration flight. Thankfully, it wasn't a serious threat, but it was nonetheless a sobering reminder of this risk. What's worse, the seeds of serious cancer could also be sown during a spaceflight, taking years to ultimately grow into a deadly disease.

There are really two ways to protect yourself from radiation. First, minimize your time in space, which reduces risk. Astronauts have to be on the ISS 365 days per year, so there's little we can do there other than limit career doses for individuals. NASA's rules limit astronauts to between one and two years of cumulative time in space, depending on the dose of radiation each individual receives. The second way is to have shielding. Our sleep stations on the ISS have some foam bricks that they say reduce radiation, and I'm sure they do. Some. But I can't imagine they help that much. The best shielding is actually water, though our living areas on the ISS don't have water surrounding them. So for now we just roll the dice.

Throughout my time at NASA, I would occasionally ask our doctors about what we knew about radiation and cancer. A few themes were common. First, astronauts die from cancer at a higher rate than the general population, which is surprising given the fact that they tend to be a healthy lot. Second, it's impossible to say whether this is statistically significant because so few astronauts have flown in space; there have been only a few hundred to date. To put that number in perspective, getting a drug approved by the Food and Drug Administration often requires tens of thousands of test subjects. The standard answer from NASA was, "We can't prove that your cancer was from space; it may just be genetic, or from other environmental factors on Earth." And they're right, you can't say with scientific rigor that astronaut cancer is caused by spaceflight. But you sure can make an anecdotal case that, if a number of otherwise healthy people—who just spent a significant amount of time in a heavily irradiated environment, full of high-energy particles that can't exist on Earth—get cancer, maybe there's some correlation.

> Although we measure radiation dosage on the station with great precision, we don't measure its effects on the human body. And for me, that is the most important question when it comes to future space exploration plans.

The most surprising medical experiment for me was one that never happened. After I returned from my second mission, having been diagnosed with carcinoma for the second time, I began to ask: Have radiation and weightlessness affected my DNA? I was just a fighter pilot and definitely not a qualified geneticist. But I figured that surely, given all the blood samples and medical tests that I had undergone, someone had checked my DNA before and after my mission. After all, a 23andMe test costs less than a hundred bucks. Nope. Nothing. Nobody checked my DNA at any point, pre-, during, or postflight, as best I can tell. I've spoken with some academic researchers at various medical schools, as well as the National Institutes of Health and the Centers for Disease Control and Prevention, who were shocked at this lack of basic genetic research, but there just hasn't been an effort to study the effects of radiation on astronaut DNA.

Again, I'm just a fighter pilot and certainly don't understand the nuances of DNA, and I know it's not as easy as a simple "before and after" test, but I do know that we have been sending astronauts and cosmonauts into space for decades, at a cost of billions of dollars, and we have lost valuable scientific data because of a lack of research in this area. Although we measure radiation dosage on the station with great precision, we don't measure its effects on the human body. And for me, that is the most important question when it comes to future space exploration plans. Not to mention my own pink body!

There was another surprise after I left the astronaut office, after Congress passed and the president signed into law the Treat Astronauts Act, in 2017. The bill's sponsor declared that it "makes sure that our brave men and women who venture into space receive support for medical issues associated with their service."

However, after leaving NASA, when I went to get follow-up treatment for my skin cancer, I was told that NASA didn't cover the cost. My doctors couldn't prove that the cancer was associated with spaceflight, and there was no coverage for medical care unless you were in line for another mission. Which was both funny and disheartening. First, providing astronauts with health care is much more than a benefit for them; it is necessary to fully understand the effects of spaceflight. The US government spends an awful lot of money to send us to space, and without gathering comprehensive medical data for the rest of an astronaut's life, any medical conclusions about those effects are based on incomplete data, and therefore invalid. The only medical contact NASA currently has with retired astronauts is through an annual physical, when we are asked if we saw any doctors during the prior year and what were the results. Of course, whatever "data" are gathered through this method are highly suspect to say the least and would never pass muster in a peer-reviewed scientific journal. To properly study the effects of space travel on the human body, you would need all medical data from all space travelers over the rest of their lifetimes.

Do astronauts tend to get Alzheimer's, or the flu, or broken bones, or schizophrenia, or whatever, more or less than the general population? Do astronauts get fewer colds than others, or do they suffer less vision and hearing loss? I don't know. And neither does anybody else. Because the only way to know that would be to provide comprehensive health care for life, ensuring that NASA got every seemingly innocuous piece of medical data. Only then could you begin to develop a set of data over time that would be statistically significant. And taxpayers would get true medical data return on their investment, rather than what they get today, which is nada. One final irony in this situation—the Russian space agency provides all of their flyers with health care for life, as do the other international partners, and they invited me to go back to Russia for health care should I need it.

Hopefully in the future, more serious "before and after" medical investigations will take place, especially as they relate to DNA and cancer, and legitimate, comprehensive, longitudinal health studies will take place to help us understand the long-term effects of space travel. Until then, I count myself blessed that I was able to spend seven months in space, contributing in a very small way to humanity's exploration of space.

TIME TRAVEL

Einstein and the Whole Relativity Thing

When I returned from my 200-day spaceflight, I spoke with a NASA physicist about the effects of the radiation exposure I had experienced. While we were talking about why I glowed in the dark (just kidding), I asked him about relativity and the impact that my space mission had on my body's clock. He told me that I had aged seven milliseconds less than the poor humans who were stuck on Earth during those 200 days, which was both amusing and a source of pride. A few months later, I was talking with a prominent Hollywood actress about this bonus time, and when I told her that I aged less during my space travels, she was amazed. She wanted to know how quickly she could go into space to slow time and the aging process.

The origins of the concept of relative time date back to the beginning of the twentieth century and a rather obscure German scientist by the name of Albert Einstein. As a young boy, he had struggled in school with middling grades. It turned out that the problem was not his intellect, but a lack of being challenged. The insights that he would eventually uncover completely reshaped our understanding of the physical universe. He audaciously claimed that many properties such as length, mass, and even time were relative, depending on your frame of reference. This was a radical idea, one that probably had Sir Isaac Newton turning over in his grave. Relativity is a concept that doesn't affect us in our day-to-day lives but is tremendously important in understanding the cosmos on an interstellar scale. And it would eventually affect me in a very personal, if small, way.

At the core of Einstein's general and special relativity theories is the concept that everything is in its own inertial reference frame, which defines how fast it is moving. For example, if you are standing still, you have a different reference frame than someone moving on a train. Also, the Earth isn't stationary, and because of its twenty-four-hour rotation period you are moving eastbound at roughly 1,700 kph at the equator, but 0 kph at the North or South Pole, with respect to the center of the Earth. Not only is the Earth rotating, but it revolves around the sun once per year, and it flies along at 106,900 kph with respect to the sun. But wait, there's more! The sun isn't standing still but is orbiting our galaxy at a comfortable 800,000 kph, taking more than 200 million Earth years to travel once around the galaxy.

You get the point—everything is in motion relative to something else. And because of that relative motion, mass, time, and size will all change relative to other inertial reference frames. Yes, this is really true. Let's assume for a minute that you're a fighter pilot, because the universe clearly revolves around you. From your point of view, you appear to be perfectly normal and your watch ticks off the seconds at a normal rate, but from your friend's point of view your time has slowed down and you've also gotten shorter and fatter. For the purpose of this chapter, let's focus on that change in time.

One of Mr. Einstein's most profound insights was that the speed of light is always constant, no matter how fast you're moving, everywhere in the universe (black holes are problematic for some laws of physics, but that's a topic for another book, one written by a proper physicist and not a fighter pilot). To illustrate this concept, consider two trains moving toward each other, each traveling 50 kph. If you are riding on one, it appears that the other train is moving at 100 kph toward you. However, if you shine a light beam at someone on the other train, you will measure its speed as it leaves your flashlight as 300,000 kilometers per second, and when that light beam hits him, he will also measure it as 300,000 kilometers per second. He won't perceive the light as traveling that extra 100 kph. No matter where you are or what speed you are traveling, light always appears to travel at the exact same speed. Even if both trains were traveling at 99 percent of the speed of light, a passenger on the other train would still measure your light beam as hitting them at the normal speed of light. Bizarre, but true.

The constant speed of light is one of the foundations of physics because it has a lot of implications for how the universe works. One of them is the fact that the speed of light is an absolute speed limit—nothing can travel faster than light. Another is the fluid nature of time. Let's say you are on Earth with your buddy, and you both synchronize your wristwatches. He goes off and flies around the galaxy at near the speed of light. When he comes back and you compare watches, yours will be much later than his. His watch will have counted off much less time than yours. He not only appears to be, but in fact is much younger than you, thanks to his high-speed galactic travels. There is a scene in the movie *Interstellar* where the crew goes down to a planet near a black hole in a very strong gravity field, which is the physics equivalent to acceleration. They feel that they are only there for a few hours, but when they return to their crewmate orbiting far from the black hole, they find that he has aged several decades, due to relativity.

What exactly causes this strange time dilation? Einstein proposed two distinct theories—special and general relativity. This is an oversimplification, but the special theory of relativity claims that time is relative based on *velocity*. However, general relativity says that time is relative based on *acceleration*, which can also be measured by the presence of a gravity field. These theories lead to the conclusion that the speed of light is constant—which may not be intuitive, but trust me, it's true.

There are several contradictory implications of Mr. Einstein's theories to my time in orbit. In my case, traveling 28,000 kph relative to the surface of the Earth, time slowed down a little bit for me compared to earthlings, thanks to the special theory of relativity. However, I was at a higher altitude (approximately 400 km), where Earth's gravity (aka acceleration) is slightly less than it is on the surface. This reduced gravity field sped time up for me relative to folks on the surface of Earth. However, during launch and landing I was under significantly increased acceleration, which slowed time down.

So, adding up all of these effects—time slowing down because of my higher velocity, speeding up because of slightly less gravity in orbit, and slowing down very briefly under launch and landing accelerations, between November 23, 2014, and June 11, 2015, I actually aged a little less than everyone reading this book (unless your name is Anton or Samantha). It was only 0.007 second according to my NASA physicist friend, but hey, I'll take it!

RE-ENTRY

PREVIOUS PAGE: Back on Earth, landing in the Kazakh Steppe. Not exactly a soft landing, but a safe one.

RIDING THE ROLLER COASTER

Re-entry Is Not for Sissies

There are several key things that every spaceship has to do if it wants to leave orbit and come back to Earth. The most obvious is changing its flight path to bend down toward the atmosphere, where the air drag will capture it and bring it relentlessly down to the surface. Next is withstanding the tremendous temperatures of re-entry. Changing your flight path angle in an airplane is a relatively easy thing; you push forward on the stick and the air pressure on the elevator moves the nose of the airplane down and the trees get bigger. Pull back on the stick and the trees get smaller.

However, in space we have Sir Isaac Newton to thank for a very useful trick that allows astronauts to come home. Orbital mechanics are what determine a spacecraft's motion once in space, and one of their implications is that to change your course to the left or right you need a tremendous amount of delta-v, or change in speed. Because of this, it's very inefficient to change your inclination, or heading. Most human spacecraft carry only enough rocket fuel to change their heading by a few tenths of a degree to the left or right. The good news is that we don't have to move left or right to come back to Earth, we just need to go down. Here's where the useful trick comes in handy—if you slow down, your orbit will descend. Conversely, speeding up makes your orbit climb. The amount of delta-v required for this trick is much less than for changing your inclination.

So, when it was time to come back to Earth in both the shuttle and Soyuz, we turned the rocket around backward, fired the engine for a few minutes, slowed down by a few hundred mph, and our orbital flight path trajectory was bent downward toward the planet. This put us on an inevitable collision course with the atmosphere and our eventual landing site, which still on the other side of the planet. While the rocket is firing it is a gentle ride, only a few tenths of a g, nothing at all like dramatic Hollywood movies with astronauts screaming and being smashed into their seats. That was launch. After the burn finished, we had some time to relax and enjoy our last few minutes of weightlessness. Because once we contacted the atmosphere, about twenty minutes later, at what we call EI (entry interface), there was no more relaxing.

It was at EI that the shuttle and Soyuz experience diverged. Dramatically. The space shuttle was a magnificent flying machine, roughly the size of an airliner, and once it was back in the atmosphere it could bank and turn and maneuver like a normal plane. Except it was traveling at 17,500 mph and was surrounded by a cocoon of plasma that was as hot as the sun, created by the indescribable friction of the massive shuttle smashing into unsuspecting O_2 and N_2 molecules of the vanishingly thin upper atmosphere.

Because we were still supersonic until a few minutes before landing, people in Florida below us heard a very distinctive, double sonic boom from the shock wave the shuttle created as it smashed into air molecules faster than they were able to get out of the way.

The view from the pilot's seat was spectacular. At first there was a gentle pink glow outside my window, then it began to glow a brighter orange and then red, accompanied by a flashing white light above the overhead window, reminding me of the scene in *Alien* when the strobe light was flashing while the ship was getting ready to self-destruct. Thankfully, this final phase of my mission took place in darkness, so I was able to see every nuance of the colorful plasma. It finally turned gray, and I raised the visor on my helmet and leaned over to the window. The plasma was slowly swirling around, like eddies and currents on a pond. I reached up, pulled my hand out of my glove, and felt the window, which surprisingly wasn't at all hot. The most bizarre thing was a very distinct yet faint sound, like tapping your fingertips gently on a counter.

Ironically, as *Endeavour* continued to slow from the mounting air pressure, things began to speed up in my brain. The airspeed felt by the shuttle's wings steadily increased, and the g loading built up to roughly one and a half g's. Because our orbital track did not take us exactly to the runway at the Kennedy Space Center, we had to make several S-turns to fly toward our destination, taking advantage of the orbiter's big wings. Our first roll reversal took place over Central America as I peeked out the window, trying to get a glimpse of the ground speeding by below, but I couldn't see anything other than a few city lights in the darkness.

As we descended, *Endeavour*'s indicated airspeed (essentially air pressure) steadily increased while her altitude and Mach number decreased (Mach 1 is the speed of sound, Mach 5 is five times the speed of sound, etc.). Because we were still supersonic until a few minutes before landing, people in Florida below us heard a very distinctive, double sonic boom from the shock wave the shuttle created as it smashed into air molecules faster than they were able to get out of the way. Once we began our final turn to line up with the runway, Zambo (George Zamka, our commander) let me fly *Endeavour* for a few minutes. As a test pilot, this was one of the highlights of my career.

The flying qualities of our rocket-turned-spaceship-turned-airplane were not great. It had what is called a harmony problem. It was very sluggish in roll, but very sensitive in pitch. It also had a quirky feature common to any delta wing airplane—if you pull back on the stick to climb, it first drops a bit in altitude, and then as the wing catches more air it finally climbs. This isn't a big deal up at high altitudes, but in the final few feet before touchdown on the runway it was a serious trap that shuttle pilots trained extensively to avoid, because a sudden command to pitch up would lead to an abrupt touchdown. My job during those few minutes of stick time was to keep us centered and on the path that computer guidance was commanding. After those brief minutes of fame, Zambo took back control of *Endeavour* for final approach and landing.

My next job as PLT (pilot) was to be a cheerleader—calling out altitudes and airspeeds as we performed what amounted to a 20-degree dive-bombing flight path on the outer glideslope to the runway. When we were 2,000 feet above touchdown, Zambo slowly pulled up to aim down the runway on the 1.5-degree inner glideslope. At 300 feet I put down the landing gear, my important task of the whole mission. Zambo greased the landing, it was perfect, and I occasionally remind him that it was the best shuttle landing I had ever experienced. Of course, it was also the only one. There was still quite a bit of piloting to do, though, as he flew the nose gear precisely down to the runway at the proper speed; getting that maneuver wrong could have led to a violent slap-down that would have cracked the fuselage. He kept our 220,000-pound vehicle on the centerline as it hurtled down the runway at nearly 200 mph, and I deployed the drag chute to slow us down. All the while a continuous stream of fire spewed from the back of the orbiter, where our rocket-fuel-powered hydraulic pumps vented their exhaust. Videos of the STS-130 landing looked like a scene from a *Mad Max* movie. As we slowed to less than 50 mph I jettisoned the chute, and shortly after that Zambo made the radio call, "Houston, *Endeavour*, wheels stop." We could finally breathe again.

I liken the experience of landing in a space shuttle to a nice, smooth Air Force landing. But let's go back to EI (400,000 feet above Earth's surface) and switch our narrative over to the Soyuz, because there are other adjectives to describe that experience. If coming back to Earth in the shuttle is like riding an airliner, being in the Soyuz is like riding a bowling ball.

The first noticeable difference was shortly after EI as we reentered the atmosphere. This time it occurred in daylight. Capsules like the Soyuz, Apollo, SpaceX Dragon, and Boeing CST-100 all use bank angle just like an airplane does to turn, though much less effectively. While the shuttle had a cross-range of more than 1,000 miles, a capsule returning from orbit can typically turn only 50 miles to the left or right. As we were zooming over Africa, we banked to the right, and when I looked out the hatch at the ground below, we were moving fast! You don't notice your speed up in orbit, 250 miles above the planet, but by this time we were only about 50 miles above the deserts and mountains, and still zooming by at several miles per second. It was so impressive that I scribbled a few unintelligible notes to myself on my kneeboard, trying to draw my fleeting view while scrunched up in that tiny capsule and bulky spacesuit.

The actual EI phase was also quite a bit different. Although I saw the same red/orange/pink glow out my window, the Soyuz was much more violent. First of all, the Soyuz separated into three parts with a giant bang minutes before EI: an empty orbital module, the descent module where we were, and an unmanned service module. After hitting the atmosphere, the external Soyuz heat blanket burned off, per design. There were constant banging and ripping noises as I watched pieces of the blanket (and who knows what else) fly by my window. Then came the parachute. We had had a briefing by crewmates who had done this before, and they basically said, "You're going to think you're going to die, but don't worry, you won't." And you know what? It felt like we were going to die. But, thanks to the briefing, Samantha, Anton, and I had a blast when the drogue chute came out. We were hooting and hollering and yelling in Russian, "Rooskiy gorkiy!" Which means "crazy roller coaster!" In the F-16 community, we would have called this phase of flight "Mr. Toad's wild ride." The tumbling lasted a few minutes until the main parachute finally deployed and we were stable and calm, back at one g.

> I had in my right hand a control stick that doesn't control anything, but gives the crew some primal comfort from the idea that they have some semblance of control, and a checklist in my lap.

Next came the waiting, as we slowly descended the remaining few thousand feet to the Kazakh Steppe. Just when things seemed to be smoothing out, the seat stroked, violently raising itself about a foot up from the bottom of the spacecraft. This stroking allowed a shock-absorber device to cushion the impact a bit. Each crewmember has his own couch, form-fitted to his body; mine had been cast about two years prior, at the Energia factory near Moscow. During that procedure, you put on white long underwear to cover all of your skin and get lowered by a crane down into wet plaster. When it finally sets they pull you out, and voilà, you have a seat liner that is molded for your body. As the Russian technicians finish this seat, they manually carve out extra room above the top of your helmet area, and I used every bit of it. On Earth I fit without a problem, but after 200 days in space I had grown a few inches and the top of my head was butted up against the top of the seat liner.

Before stroking, I didn't have much room in the cockpit. We were all in our bulky and uncomfortable spacesuits, wedged into a volume that was

A few seconds before impact in Kazakhstan—me, Anton, and Samantha strapped into our Soyuz TMA-15M.

roughly the volume of the front seat of your car, with small pieces of equipment wedged into every inch of free space. After stroking, I was moved up so that there was probably a foot between the control panel and my face. My right arm was smashed against the capsule wall. My knees were in my chest—no stretching your legs, there's a capsule wall in your way. I was strapped down very tightly so I couldn't move. I had in my right hand a control stick that doesn't control anything, but gives the crew some primal comfort from the idea that they have some semblance of control, and a checklist in my lap. I thought to myself, "OK, I'm not claustrophobic, but if there was ever a reason in my life to panic it would be now." I figured I had two choices: a) panic, in which case I'd be strapped in, unable to move, with absolutely nothing to do about it, or b) not panic, in which case I'd be strapped in, unable to move, with absolutely nothing to do about it. I chose option b.

The last few minutes before touchdown were spent quietly waiting. Hands on my checklist, which was on my chest. Controlled breathing. Making sure my tongue wasn't sticking out between my teeth because I didn't want it shortened upon impact. Watching but not trusting the altimeter because it can have errors of several hundred meters. There was nothing more for our commander Anton to do; we were floating helplessly under the parachute, waiting for touchdown. The Russian Air Force was there waiting for us in their Mi-8 Hip helicopters; they got a visual very early, calling out our altitude over the radio.

Then everything happened at once. A loud tone, explosion, violent crash, seemingly bouncing out of my seat, being thrown sideways. The Soyuz has "soft-landing" rockets on the bottom of the capsule, designed to fire a split second before impact, but my suggestion is to rename them "less-of-a-crash-landing" rockets, because a crash is exactly what it felt like. I imagine driving into a telephone pole in your neighborhood would feel roughly like a Soyuz landing. But the combination of form-fitted seats, soft landing rockets, and shock absorbers in the seats made the landing entirely safe, with the exception of a few minor bruises. Shortly after we landed and rolled 360 degrees, back to an upright position, someone on our crew said, "Are we alive?" The three of us put our hands together—we had survived and were back on our home planet!

Getting out of the space shuttle had some similarities to getting out of the Soyuz. As *Endeavour*'s pilot, I had a lot of tasks to accomplish postlanding: doing a detailed checkout of the flight control and hydraulic systems, powering down the thrusters, etc. Which meant I was the last person to get out of the shuttle. That became a trend; five years later I was also the last out of our Soyuz, and even now when I'm in a crowded car or airplane I tend to be the last one out.

Launching into space, accelerating from 0 to 17,500 mph, riding a rocket trailing flames, shaking and roaring and smashing you into your seat for the eight-and-a-half-minute ride to orbit is a kick-ass experience unlike anything you'll find on Earth. But the ride back to Earth, slowly decelerating from 17,500 to 0 mph, is even more amazing. Many countries have launched rockets, but only a handful have successfully brought people back home from space, and there's a reason for that. Re-entry is hard. It's an incredible experience, but a dangerous one. And if you ever get a chance to do it, you'll feel like you're going to die, but trust me, you'll be OK. . . .

ADAPTING TO EARTH

You Try Walking After Six Months in Zero G

Although getting into space—accelerating from 0 to 17,500 mph—is an incredibly difficult task, coming home from space—decelerating from 17,500 to 0 mph—is even more difficult. The technical intricacies of thermal protection, attitude control, and guidance and navigation make it very difficult to come back to Earth. I knew this before I ever flew, at least from a technical point of view. But I was also concerned from physiological and spiritual points of view. How would my body readapt to gravity? Would I miss space? Would I feel out of place in mundane daily life after the sublime experience of space? Would I drop things absentmindedly because they don't float anymore?

My first spaceflight was a two-week experience. But it was long enough to affect the whole crew with a sensation of both dizziness and feeling heavy. Adjusting to gravity began while we were still in the middle of re-entry. Decelerating through Mach 20 (twenty times the speed of sound, or about four miles per second), there was a fair amount of acceleration and force on the shuttle from the atmosphere smashing into *Endeavour*'s heat shield. We were finally experiencing weight again, and with that our brains' vestibular systems were asking, "What the heck is going on here!?" They were probably much less polite, because it was extremely disorienting!

One of the lessons we had learned from astronauts who had gone before us was a technique to get our brains readapted to gravity quickly, so we would not be too dizzy for landing. During the re-entry phase, after the onset of the effects of gravity but about twenty minutes before landing, Zambo and I slowly rotated our heads up and down and left and right. Forcing our brains to deal with the "new" sensation of gravity, in multiple axes, by rotating up/down/left/right caused our brains to adapt as quickly as possible. As the two pilots on board, it was critical that we both be able to land the shuttle with minimum disorientation. I'm sure nodding and shaking our heads slowly looked strange, but we weren't worried about looking cool; we only wanted to be ready for landing. It worked for me. I did not experience any problems with disorientation, and based on his perfect landing I don't think Zambo did either. Some of our other crewmates had a problem with disorientation after landing, and I wonder if doing our technique would have helped them with that.

Another acute problem during re-entry was dehydration, caused by body fluids pooling in our legs—which could potentially lead to orthostatic intolerance, that light-headed feeling you get when you suddenly stand up. When you first get to space you have to pee a lot, reducing the total amount of fluids in your body. That isn't a problem in weightlessness, but it is down here in the land of gravity. To combat that, we had a very elaborate system of fluid-loading. One of the mission specialists helped us by managing a series of drinks that we had to take every fifteen minutes, beginning an hour and a half before landing. We could choose from water with salt tablets, Powerade, chicken broth, or any kind of salty water. It wasn't tasty, but it was important.

My fluid-loading program had a slight hitch—my buddy was telling us, "OK guys, time to take your next drink bag" at fifteen-minute intervals, but when everyone else was done he still had me drinking more. I finally got to the point that I was about to pop—and keep in mind, we were in our pumpkin suits, strapped in for re-entry, with only a diaper to use. I cried uncle, went about our normal descent profile, and landed. When I was back on Earth I felt very good, no light-headedness or sick feelings. And then I learned that my comrade had accidentally given me the wrong fluid-loading prescription, basically double what I needed. I say "accidentally" because he may or may not have been getting me back for a certain prank I played on him on the launch pad, but it was OK with me because I felt so well once I got back to Earth—after finding a bathroom. So my advice to you the next time you need to return to our planet from space—do lots of fluid-loading.

After landing the shuttle, Zambo and I had a lot of procedures to perform as commander and pilot. Communication with mission control, safing the hydrazine-powered hydraulic pumps and rocket propellant systems, getting the computers and electrical systems in a good configuration, etc. While we were busy shutting down *Endeavour*, our crewmates were getting out of the vehicle, one by one. When it finally came my turn, I grabbed my helmet and tried to hand it to the ground crew, telling him, "Be careful, this thing weighs five hundred pounds!" This was a perfect description of the next few hours and days for me—everything felt heavy! When I finally got back to my room late that night, it felt like my blanket was made of lead, and I was a superhero who was pinned against the bed by a villain with a giant magnetic force field. A day later, when I was finally home, I took my son out in the driveway and we played some basketball. Except whenever I took a shot I couldn't even get the ball up to the rim, much less in the hoop. I swear someone filled our basketball with lead during my two weeks off Earth—maybe that same mission specialist who was helping me with fluid-loading. I wanted to tell President Obama this story because he is a basketball fan like myself, but it slipped my mind when I had a chance to meet him a few weeks later.

Feeling heavy wasn't the only issue; there was also dizziness. About an hour after landing, we were out of our orange spacesuits and into our more comfortable blue flight suits, and after a medical checkup we were on the runway underneath *Endeavour* performing a walk-around, a basic postflight inspection that every pilot does after landing. Wow, was I dizzy. I was able to do the whole walk-around, but I wanted someone next to me, just in case I tumbled over. Luckily I didn't, but it was a fairly unpleasant experience. I also noticed that although a few of us felt sick when going into space (I was among them), different people felt badly upon return to Earth (not me, thanks to fluid-loading). My flight surgeon always told me that you couldn't predict who would get sick going uphill and who would get sick coming back to Earth; it was a bit of a random proposition. And based on my limited experience I can confirm he was right.

I don't know why this is such a common question, but it is—"Did you drop anything after you got back to Earth?" The answer for me was a resounding *yes*. The morning after my shuttle landing, I was at a local hotel in Cocoa Beach with my family. I was holding a water bottle, and when I went to give it to someone, I floated it to them in a straight line, except, gravity. It fell straight to the ground. It was an instinct; my brain had gotten used to floating things in a straight line. Unfortunately, that doesn't work here on Earth. . . .

Things were similar after my 200-day mission, but the heavy feeling was less intense and the dizziness was more intense. That first day was painful. It was like I had had a bottle or two of wine. I was able to walk and move around, but I hated it. NASA invented a torturous test in which we were required to lie on our stomachs and stand up as quickly as possible to test for orthostatic intolerance. They also made me close my eyes and walk, one foot in front of the other, toes touching heels, in a straight line. It's hard enough to do that right now, but try doing it after nearly seven months in space! I made it through all of that torture and could walk around on my own, but I really wanted either a person or handrail next to me that first day.

Actually, what I really wanted was a bed. Our Soyuz capsule had thumped down on the Kazakh Steppe and rolled over 360 degrees, coming to rest upright. It's much better when it comes to a rest on its side, as it usually does, because then the big burly Russian ground crew can crawl into the Soyuz and drag you out and plop you down into a La-Z-Boy, set up right there in the desert. However, in the rare case when the Soyuz comes to rest upright (like ours), we have to crawl out the top hatch on our own. Then the big Russian guy grabbed us and put us in the lounge chair. You rest for about thirty minutes, call home on the satphone, and then go to do medical testing. That was one of the most humiliating things about my whole flight. We were in the middle of nowhere, so there was no bathroom. I got out of my spacesuit, pulled my long underwear down, and peed into a bottle while my two flight surgeons held me steady by each arm. That was a first for me, being held upright by two medical doctors while I peed. They then gave me a preemptive IV to get some fluids in my body in order to stay ahead of dehydration, just as fluid-loading does. Next was a helicopter ride to the airport in an old Soviet Mi-8 Hip as the sun was setting—it was beautiful!

After a twenty-four-hour trip back to Houston on a NASA Gulfstream business jet, modified with special beds for deconditioned astronauts, I went straight to the astronaut gym to begin my rehabilitation program. There,

Bruce Nieschwitz, my ASCR (the NASA acronym for strength and conditioning specialist), ran me ragged with all of the tasks I described in chapter 11. A week after I landed, I was nearly back to full strength, thanks to Bruce's workouts. I went in and did rehab every day, even on weekends. I strongly believe that exercising every day while in space and then dutifully going to rehab every day once back on Earth is the key to minimizing bone and muscle loss. I only skipped four days of exercise while on the station—the three spacewalk days and the ammonia leak day we spent in the Russian segment thinking the station was going to die.

In addition to physical strength, NASA also measured our balance. Before launch I had done a series of neuro-vestibular evaluations, standing in a movable box (think Halloween fun house), in a harness to prevent a fall, vision impaired so I could only see the moving wall in front of me. The researchers would abruptly jerk the box and measure the force pattern that my foot made as I attempted to remain standing and not fall. It was really annoying—like having someone pull a rug out from under you. If your foot was pressing erratically on the front and back and side, quickly changing its force distribution, it meant that your sense of balance wasn't good. If the force pattern from your foot made a quick adjustment after the box shifted and quickly stabilized, it meant that your inner ear and brain and all of its balance software were working well. After each session, I was given a numerical score to evaluate my balance.

It turns out that I have pretty good balance in general. But what absolutely shocked me was the test I did one week after landing. I scored better than I had before launch, nearly seven months earlier! The scientists had never seen that before, and I was personally shocked, but the data were there. My body readjusted to Earth extremely quickly. The first day was very painful, the next a little less, the next even less, and a week after coming home from a 200-day mission my balance score was better than it had been preflight. I was blessed with a body that was made to go to and from space.

Beyond the physical is the psychological aspect of leaving and returning to Earth. This had me concerned. Would I miss my time in space? Would I feel out of place back home, in a predictable universe of meetings and kid activities and traffic and bills? Would I be depressed, as other astronauts had been? I looked at spaceflight like winning a Super Bowl or World Series or Academy Award. It's an amazing thing, something that may happen only once or a few times if you're really lucky, and then after it's over, what do you do? The answer to my concern came to me on my second day on Earth. I had landed in Kazakhstan, made the twenty-four-hour trek back to Houston, and done my rehab workout and finally was set free. My son, who had turned sixteen and gotten his driver's license while I was in space, said, "Dad, let's go car shopping." So I got in the car with him (he drove) and we went to the local Ford dealer. I vividly remember the first red light where we stopped, as I thought, "Well, I'm back on Earth. It's like a switch in my brain just got moved from the 'in space' position to the 'on Earth' position, and it feels fine! Earth is 100 percent normal and I don't miss space at all; I'm just back on the planet and it's great."

Since that moment, I haven't looked back. Flying in space was absolutely incredible, a blessing beyond belief. And now I'm back on Earth, and it's awesome. We have a beautiful planet and some great people, along with plenty of problems, but it's our planet. Now that I'm home I want to live my life with purpose, looking to the future, not stuck in the glory days of the past, and thankfully that's exactly how my mental state has been since I came back down to Earth.

TRAGEDY

Being There for the *Columbia* Catastrophe

This chapter is hard to write. The others have flowed onto the laptop; I didn't even have to think as the words just moved from my brain to my computer. But this is one of the final chapters I'm writing for this book because it is the one I least wanted to write. It's the one that hurts the most, but it is nonetheless a part of human spaceflight. Adventure comes at a cost, and sometimes adventurers do not make it home. This is my personal story of dealing with a space tragedy.

We were waiting by the side of NASA's Shuttle Landing Facility (SLF) at the Kennedy Space Center, where *Columbia* was supposed to land on Runway 33, on the morning of February 1, 2003. It was a bright, sunny Florida winter day, and everyone was exhausted. This mission had been planned for years, slipping time after time as the space shuttle program experienced technical problem after technical problem. Because STS-107 was purely a science mission, its priority was lower than that of the politically sensitive, space station assembly flights. This was to be the final shuttle flight not dedicated to the space station or Hubble Space Telescope, and her crew of one Israeli and six American astronauts had done a remarkable job performing that science during their mission. Today was the seventeenth day since launch on January 16, and everyone was ready for them to come home. The families were tired after so much stress and excitement and lost sleep, NASA's flight control team was ready to bring them home after several weeks of nonstop work at mission control, and I'm sure the crew was ready as well.

There we waited in the February morning sun, a group of about fifty family members, astronauts, and NASA support personnel, shielded from the prying eyes of the press. NASA had learned from the *Challenger* accident in 1986, when the shuttle exploded shortly after liftoff and family members, who were colocated with the press, were photographed in that moment of horror. There was now a detailed contingency action plan to keep families carefully sequestered during launch and landing, with specific steps to take in the event of disaster. As an official family escort, I knew I would be an integral part of any such plan. I had run across countless checklists and procedures during my time at NASA that were required reading: how to evacuate the building in the event of tornado, what to do if a hurricane hit Houston, etc. But this contingency action plan was one I took seriously.

We were chatting with each other, small talk about the kids having to get back to school on Monday (it was a Saturday), what plumbing or computers had broken at home during the mission, gossip about who in the astronaut office was getting assigned to the next shuttle flights, who was sick, etc. Normal adult conversation while waiting for a space shuttle to land. NASA had piped in audio from mission control to a loudspeaker, and the normal chatter of crew-to-ground conversations, as well as the NASA TV description of landing, filled the air. About fifteen minutes before landing, *Columbia*'s crew stopped talking, and the CAPCOM, astronaut Charlie Hobaugh, made several "*Columbia*, Houston, comm check . . ." radio calls. No reply. I didn't notice immediately because we were busy with small talk, but one of the family members came over to me and asked, "Hey Terry, what's going on, there's no comm?" I absently said that this was normal, the extreme temperatures during re-entry could block the radio signal to the shuttle, especially as it went through a series of S-turns, banking left and right to fly toward the landing site. I assured him it was OK, and we'd hear from them soon and went back to my chitchat.

For a few moments. About two minutes later a hush came over the whole crowd, as the oppressive sound of silence on the loudspeaker dominated everything. I very quickly ran through some possible scenarios in my head. This could be a true communication system failure. But I knew how much redundancy there was—*Columbia* had radio links to communication satellites as well as antennae on the ground. Two UHF radios, just like in military jets, to talk to the ground. An extensive telemetry system that would provide mission control with data, even if there were no voice comm. I knew something really bad had happened by L-12 (twelve minutes before landing). We had no television or radio or news of any kind, and thank God these were the days before iPhones and Twitter. So we sat there, in the blind, waiting.

By the time the giant clock had counted down to L-3 minutes, we all knew we should have heard the twin sonic booms from the orbiter flying overhead, still supersonic 50,000 feet up. But there was nothing but deafening silence. I knew the ground tracking radars and cameras would have zoomed in on *Columbia* by now. And I knew she wasn't coming back to the SLF that day. Jerry Ross, one of our most senior astronauts, directed us to get the families into our rental vehicles and drive them to crew quarters, the astronaut hotel at KSC where we slept before and after missions, got suited up before launch, and were reunited with our families after a mission. We would all wait there for news. As the clock approached L-00:00:00 I knew what had happened. I had an overwhelming sense of sympathy for the people who had said there was nothing to worry about when a large piece of foam had hit *Columbia*'s wing eighty-one seconds after liftoff, seventeen days ago. My brain had processed that event, and I immediately, intuitively, knew that was the cause of a grave accident, long before I knew any of the actual details or had seen the tragic images of *Columbia*'s debris streaking high in the Texas sky.

Several of the family members got in my SUV and we drove in silence, the longest fifteen minutes of my life, car radio off. I didn't want them to hear a news report. My cell phone rang after a few minutes—it was my wife, who was crying, having seen the news on TV. I calmly said, "Hi, yes, we heard there is no comm with *Columbia*, and we are all driving to crew quarters. We don't know any details yet, but I'll let you know as soon as we do." And quickly hung up, acting as though it was still a mystery. But from that brief call, I knew there was no mystery. Within minutes of arriving, Bob Cabana, another senior astronaut, came in to crew quarters and broke the news to the families and their astronaut escorts. *Columbia* was lost along with her crew. There were no survivors.

My time as a family escort for STS-107 had begun several months before, when I heard that Rick Husband wanted to talk to me. I was a rookie, one of the new guys, and Rick was our senior Air Force astronaut and a shuttle

commander. He was a *big* deal, and I was a nobody. So I immediately assumed I was in trouble when I heard I was to go talk with him. Much to my relief, he greeted me with his ever-present big Texas smile and was glad to see me. He wanted me to be one of his crew's four family escorts. Most shuttle crews had only two, but because STS-107 had the first Israeli astronaut, Ilan Ramon, their crew would have extra support to help with the added international scrutiny. I was blown away by this huge honor, one normally reserved for more senior astronauts. I immediately said yes and excitedly went home to share the big news. I felt this was one of the most important jobs of my life. I had no idea just how true that would be.

I spent the next few months getting to know the families of the astronauts. My job would be to help them through all the major milestones of a shuttle flight: going to launch, escorting them into mission control for family video conferences (think space-Skype), keeping them up-to-date with the progress of the

And though it was never verbalized, everyone close to me was painfully aware of the dark side of spaceflight. The risk was real, and the consequences of failure were even more real to my family than most.

mission, and finally going back to Florida for landing. Not to mention helping kids go about their routines of school and activities, organizing launch and landing parties, and other mundane daily life things. I looked at this as a way to do whatever they needed while their astronaut was in space. I knew my family would appreciate the same support when it was my time to fly.

The *Columbia* crew was special. Really special. First of all, they were all very good at their jobs. I sat in as an observer during one of their ascent simulations, the hardest and most challenging part of being an astronaut. I had thought that flying single-seat F-16s on night attack missions was the busiest a human being could be, until I trained as a shuttle pilot during ascent simulations. Watching Rick in the left seat as commander and Willie McCool in the right seat as pilot, along with Dave Brown and Kalpana Chawla as mission specialists, all working together during a constant stream of diabolical malfunctions dreamed up by the simulation supervisor, blew me away. As a pilot, or right seater, it seemed that I had no hope of ever being as good as Willie.

He handled every emergency situation perfectly, communicated with Rick and mission control clearly, and seemed to be a robot, incapable of making a mistake. Seeing such a great crew was intimidating to say the least, and it set a bar for me to strive toward.

As Israel's first astronaut, Ilan Ramon was a true national hero. In a country marred by continuous strife, Ilan was a rare source of hope and unifying force within Israel. He had an air about him that made me think that he was capable of anything; in a room of a thousand people, he was the one who would rise above the crowd and everyone would trust and look to as leader. A decorated F-16 pilot, Ilan had been number 8 in the formation of eight F-16s that had bombed Iraq's nuclear reactor back in 1981. Being the last guy is always the most dangerous position in a formation, because by the time you flew over the target, all enemy antiaircraft guns would be awake and trained on you. Our World War II bomber pilots called this dubious distinction "tail-end Charlie." He survived that mission over Iraq, sparking a legend that would only grow in his home country. More important than any Air Force or space success was his family. His beautiful and wonderful wife, Rona, meant the world to Ilan, and his four children were all special. I could tell, even at their young age, that they would all grow up to be impressive adults and leave a mark on the world, as their father already had. It was an honor and pleasure to serve him and his family.

The rest of the crew was equally impressive. Laurel Clark was a Navy doctor who always had a smile on her face. Dave Brown was a bit of an underachiever—Navy fighter pilot, medical doctor, aircraft owner, he made me wonder why I had wasted my life away! Mike Anderson was a spaceflight veteran, fellow Air Force pilot, and payload commander, in charge of all science experiments on this NASA mission devoted entirely to science. His lovely wife would become a lifelong friend of mine. Kalpana Chawla had been born in India, emigrating to the USA when she was twelve. She was an amazing engineer and the kindest human I have ever known. Rick was a natural-born leader, full of West Texas wisdom and anecdotes ("You can't swing a dead cat in here without hitting . . ." was one of my favorites), and his family became an extended family for me over the decade following the accident. Finally, Willie

McCool, one of the smartest and most capable humans I have known. His wife, Lani, and I have become lifelong friends; she is an artist at heart and an incredible photographer, and a mentor to me. She has given me wisdom and helped me to see things in a new light as life has taken twists and turns that neither of us could have imagined back in the winter of 2003.

The astronaut office is full of smart people. But the STS-107 crew was special—more than just smart, they were genuinely decent and remarkable humans, the kind of people I wanted to grow up to be like. It was an honor to serve them and a tragedy to lose them.

Which is part of the reason why it hurt so badly. In the ensuing weeks and months and years, I never heard of any NASA manager or engineer who was deliberately at fault in this accident. The shuttle program was full of dedicated and smart people, and frankly this disaster hurt those who were directly responsible. Which is why I immediately felt bad for whoever had approved moving ahead after the "foam strike" that would ultimately kill the crew.

In fact, I was to blame as much as anyone else at NASA. As family escort, I was getting a steady stream of mission updates, and a few days after launch an email came to me with a short video clip of the foam popping off the shuttle's fuel tank into the 500-knot windstream and getting blasted back into *Columbia*'s wing, causing a split-second explosion as the shuttle continued its climb into orbit. I was immediately concerned and walked down the hall to a more senior astronaut, asking him if we could take some pictures of *Columbia*'s wing to see if there were any damage. I was no rocket scientist, but I knew that heating during re-entry was critical and that the leading edge of the wing was a particularly critical area, having a special heat shield to withstand the fiery plasma. I was told that they had looked at it and deemed it not to be a safety-of-flight issue. Besides, what could be done if there was damage? There were no repair kits on board, and the next shuttle wouldn't be ready for at least a month even if they decided to launch a rescue mission. The crew would be OK, and besides, NASA didn't want to bother the Air Force to ask them to take photos of any potential damage. I was actually told that by a senior astronaut. It didn't make sense to me; I thought it would be much more prudent to get imagery of any damage, and if a serious problem were found at least we could try something, but I was a rookie at the bottom of the totem pole. I figured that those in charge must have known more than I did, since after all they had been flying shuttles for more than twenty years.

I said OK and did nothing. Life went on for the next two weeks as I was bringing family members to mission control, taking kids out for pizza, forwarding mission update emails, and planning for landing. But that foam strike lingered in the back of my mind. I didn't tell anyone about it because I didn't want to raise concerns unnecessarily. But it was there, gnawing at my conscience. And as I stood on the tarmac watching that landing clock count down to 0, with no shuttle in sight, I knew exactly what had happened.

At the end of the day, the problems that killed both the *Columbia* and *Challenger* crews were managerial, not technical. Yes, you could trace the accident to very specific problems: O-ring temperature limits on STS-51L or cryo-pumping foam insulation on STS-107. But at the heart of both issues was a management culture that was arrogant, that thought they knew things when they didn't, that didn't listen to those at the bottom of the chain of command, and that was more worried about external factors like keeping Congress happy with flight rates or not bothering the Air Force with crew safety. In the years immediately following both accidents, NASA had a rebirth of sorts, with a lot of soul-searching and a very healthy focus on safety. Unfortunately, as the years and decades go on, those lessons tend to be lost. I pray that this does not happen in the future.

I learned some painful lessons from this experience. If you see a problem, speak up. Even if you are junior, you need to make sure the bosses are aware if something is seriously awry. Also, just because things have been going well doesn't mean that you have been making the right decision—you may have simply been getting lucky. NASA had launched shuttles with foam falling off for twenty years and never fully addressed the problem, because they had never lost a crew. As a leader, have a culture of debriefing where rank doesn't matter. The list goes on, with many subjects for a different book.

The *Columbia* incident really hit home for me when it finally was time for my first ride into space, seven years later. My own family had experienced the STS-107 accident right along with me; they were there when I got home from work after a day of family-escort-related duties during the mission, and after attending yet another funeral or dedication ceremony in the weeks and months and years after the accident. My son and I went on a father/son camping trip with one of the *Columbia* crew's sons years later. And though it was never verbalized, everyone close to me was painfully aware of the dark side of spaceflight. The risk was real, and the consequences of failure were even more real to my family than most. This is something that is rarely discussed or even acknowledged publicly, but I can tell you firsthand, the stress is real and it takes a toll on all involved—least on the astronaut and most on the family. It did in my case, anyway.

With all of this in mind, I took some private time on the evening before my first launch and closed the door to my room at astronaut crew quarters, got out a pen and paper, and began to write. And weep. A final letter to my children and my wife, in case I died on *Endeavour* during STS-130. That was hard. Feeling those kinds of emotions wasn't usual for me as a fighter pilot and I rarely verbalized them, a problem I still have to this day. It's a bad trait when it comes to intimate human relations, but a good one when you are flying fighters into combat, or rockets into space. One of the harsh realities of my chosen profession. I knew this letter would be something treasured for the rest of their lives should the worst come to pass. Thank God it didn't.

I recently heard a fellow astronaut, full of bravado, say that "we don't ride into space with our fingers crossed." His point was that astronauts are all so well prepared that we know everything that could possibly go wrong and are ready to deal with any contingency. Of course, I knew better. There were a thousand things that might have killed me during launch and they were all beyond my control: a leaky fuel line, a hole in a combustion chamber, a bad weld in a critical structural joint. Most of these hazards would probably spiral out of control in fractions of a second, with nothing that anyone could do. The reality was and is and will always be that when you strap on a rocket, there is a chance that something could go wrong and kill you. With absolutely nothing you could do about it.

It is a risk that astronauts choose to assume themselves, but their families bear the brunt. When the global financial crisis of 2008–2009 hit, it was labeled a "moral hazard," where society had to assume the risk and therefore pay the cost of bad Wall Street decisions. In accounting you could call it an improper matching of revenue and expenses. When riding rockets, it meant that astronauts got all the fun while their families worried and wondered if they would ever see them again.

I sat there alone, weeping, letting all of my emotion out while I had this final chance, hours before my launch, to write what could become the last communication from me to those closest to me. After the envelopes were sealed and placed in an obvious place where my CACO (NASA acronym for the astronaut assigned to help your family in case you died) would find them. Then it was time to go down to the robotics simulator and do one final practice run, removing Node 3 from the payload bay of *Endeavour* and attaching it to the ISS, before I would be doing that for real in five days' time. Such was spaceflight. Moments of intense emotion, followed quickly by a reality check and getting back to work.

Exploration is dangerous, and space is completely unforgiving. You need to do things right, you need to do things smartly, you need to be humble, and you need to be lucky if you are going to survive. At the end of the day, space travel is a very human endeavor. One that is full of real people with real families—spouses, kids, parents, siblings, friends. It is risky, and as an astronaut you need to do the cost-benefit analysis to be sure that it is worth it. Because you will be subjecting your loved ones to a very real and painful risk if things don't go well.

We miss you, *Columbia* crew. The world is a worse place without you. I hope we learn the lessons that transcend borders from the lives you led. I hope that NASA remembers the lessons it learned from your loss. And I am thankful to God that my letters never had to be read.

NO BUCKS, NO BUCK ROGERS

Meeting with Washington Politicians After a Spaceflight

There is a long-standing tradition of astronauts making the rounds in DC after their mission. John Glenn famously addressed a joint session of Congress after becoming the first American in orbit, and there is a great photo of the *Apollo 11* crew in quarantine inside an Airstream trailer speaking with President Nixon. The connection between American politicians and astronauts is one that began in the early days of the space program and has continued to this day. For all of the political rancor and disagreement and divisiveness in DC, one constant remains true: Just like the timeless Axe commercial, nothing beats an astronaut.

Nowadays there are two flavors of postflight Washington visits that astronauts take after their spaceflight. The first kind is the White House visit, which depends on the president. Some presidents are really into space and invite nearly every space crew to visit, and others don't. In 2010, President Obama invited our STS-130 crew to the Oval Office; it was one of several times I have been to the White House and a highlight of my time at NASA.

There seems to be a perennial battle between astronauts and NASA HQ about what to wear for these visits. Astronauts inevitably want to wear business attire, but NASA wants us to wear the "blue suit," the easily identifiable NASA-blue jumpsuit we wear when flying T-38s and also when doing public speaking. I was firmly on the side of business attire for events as important as the White House, but every crew had to fight its own battle when it was time to head to DC. There was a legendary story from a few years back, when NASA would routinely force astronauts to wear the blue flight suit. Soon after George W. Bush was elected, he invited the next shuttle crew for a visit, and HQ immediately beat them into submission by forcing them to wear the blue suit. When they showed up in the Oval Office wearing their jumpsuits the president immediately took note and expressed his disapproval, saying, "We normally wear a coat and tie in this office," as the whole crew shrank in embarrassment, glaring at their NASA HQ escort. Personally, I have to agree with W on that one—a flight suit is too dressed down for the Oval Office.

I was very proud of Zambo, our STS-130 commander, when he held off the NASA HQ assault and got approval for our crew to wear suits and ties. The big day began at the White House security office, and when we showed up a very famous Army general was waiting in line. I was honestly not a fan of his, because of his public politics. We were whisked through security while he was stuck there being hassled and delayed, and I had a good chuckle over that. Next, we were given a very brief tour of the White House, where we saw a large table with a television set in one of the conference rooms and I immediately recognized it—it was the same room where we had done a Skype call to Mr. Obama a few weeks earlier, from space!

Finally, at the appointed time a guy in a suit wearing sunglasses (I wasn't sure why, because it wasn't that bright inside the White House) with earbuds and a bulge in his sport coat came out and called us into the Oval Office. I always chuckled at those Secret Service guys, because they all look just like their *Men in Black* counterparts, "Agent J, Agent K." The meeting itself was exactly what you'd expect: Our crew stood, shook hands, and thanked the president for the visit, then he smiled and took some photos and asked us questions. When he asked, "What was it like to come back to Earth?" Zambo said, "Terry, why don't you answer that one?" I told him about being dizzy and feeling really heavy, the standard stuff. Afterward I kicked myself, because I meant to tell him the story of how a day after landing I was in the backyard playing basketball with my son, and I just felt so heavy; I would grunt and heave with all my effort but just couldn't get the ball up to the rim. It was comical, and I think that as a basketball fan Mr. Obama would have appreciated that story. Oh well, next time.

Following the White House visit we took a trip to the Hill, where we met with twenty senators and congressmen. Before being cleared to meet with the big guys and gals, we reported to the Legislative Affairs office at NASA HQ to get our briefing. It was a very controversial time in human spaceflight because just a few weeks prior Charlie Bolden, the NASA administrator, had canceled the Constellation program, President Bush's plan to go back to the Moon and eventually on to Mars. It had been clear since Inauguration Day in 2009 that there would be some big changes at NASA, but this had been a huge change, and frankly a serious setback. The space shuttle program was drawing to a close and now its follow-on program was also being canceled, leaving us entirely dependent on the Russian Soyuz to get to the ISS, and without a vision or plan to go beyond the ISS. It was as if I were watching a train speeding toward a bridge that was out, in slow motion, and couldn't do anything about it. Sure enough, it took a decade before we once again launched astronauts from US soil, and though we are trying to jump-start a Moon program, it will be much less capable than Constellation would have been.

Given that backdrop, I knew I would be doing a lot of tongue-biting when visiting the Hill. Our Legislative Affairs escorts were nice enough, and they themselves understood the disaster that had befallen NASA, but they had to toe the party line and try to give us something coherent to say. Our crew decided it would be best to just talk about our mission to the space station and avoid the mess that NASA's human spaceflight policy had become. Off to the Hill we marched, bouncing between senators and representatives, Republicans and Democrats, members with NASA centers in their districts and those who had no idea what NASA stood for. Despite the gigantic Constellation-cancellation elephant in the room, it was a great visit—with headquarters handlers looking over our shoulders. They were good guys, but it felt like having a Soviet political officer hovering there to ensure we remained ideologically pure. It was kind of funny. Kind of.

When we would visit a Republican who understood space policy, they would say, "Look, I know you can't say what you're thinking, but this whole Constellation debacle is a mess, we will be depending on the Russians, we don't have a plan for the future, etc., etc., etc." There was a lot of ranting and raving, and I silently agreed with them. I was not alone in my extreme disappointment. When we visited Democrats, they were less up front in expressing their frustration, but the ones who had NASA centers in their districts were clearly annoyed. There was a third category—the politicians who had no interest in NASA. There was no space bacon for them to bring home to get them reelected. These were friendly, polite meetings; we would smile, get our picture taken, give them our crew montage as a gift, and move on, scratching our heads. "Why did we just visit that person? Those are ten minutes of our lives that none of us will ever get back."

Speaking of photos—Zambo had worked out a genius compromise regarding the blue suits. After being pummeled with requests to wear the full flight suit, he negotiated that we would bring a blue jacket to put over our business shirts for photos. This allowed the congressional members to get a photo op with all-American-hero astronauts in blue NASA attire, and we were able to take the flight jackets off and get back to suit-and-tie as soon as we were done. Win-win. Zambo should be in Congress, in my opinion. That was surely one of the only compromises that has happened in Washington in recent decades.

Fast-forward five years to my next mission, Expedition 42/43. I once again traveled to Washington for a Hill postflight visit, along with my European Space Agency/Italian crewmate Samantha Cristoforetti. For some reason, NASA did not invite our third crewmember, Russian cosmonaut Anton Shkaplerov, which was a bit rude. We also were not invited to the White House on this trip; those visits had mostly dried up with the end of the shuttle program unless there was something unique about the crew. So off to the Hill we tromped, Samantha and myself, into an environment where the 2016 election was front and center.

This time I was the commander, and we were a much smaller group. I was actually alone for some of the visits, so I was much more open and forthright with my opinions about space policy, opinions grounded in reality and not ideology. I spoke my mind, which the senators and congressmen seemed to appreciate and NASA did not balk at. One of the key points I made was that space policy was primarily limited not by rocket science, but rather by political science. By that I meant that we can't massively change space policy every four or eight years because the next president hates the last president and wants to pursue a different ideology. Space is hard and it doesn't care about political ideology, it only cares that $F = m \times a$. Every time I used the "rocket science vs. political science" line, I heard resounding agreement. One member lit up and turned to his aide and said, "Write that down," and I was actually quoted in his committee hearing the following week.

Every single member, without exception, agreed with me, saying, "You are so right, and if it wasn't for the folks on the other side of the aisle we could be getting this or that done." I would have chuckled to myself had this not been such a tragic commentary on our twenty-first-century political system. I have since concluded that our two-party system is broken and badly in need of a third, centrist party, one that represents the views of most Americans. The issues affected by partisan dysfunction are incredibly serious and dwarf space policy. We need to make some fundamental changes to our politics, and soon. The subject of another book.

I have had several other opportunities to visit the White House in my capacity as an astronaut. After the *Columbia* accident in 2003, President Bush invited the families of the STS-107 crew to visit him in Washington, and I traveled with them as one of their family escorts. We had all been briefed to expect a very quick visit; the families would take their pictures with the president, they would exchange a few words, and that would be it. When we arrived, both President and Mrs. Bush were there to meet us, and they invited everyone into the Oval Office for a group photo. There must have been thirty people, between family members, astronaut escorts, and NASA HQ folks. Next, they led us all on a tour of the entire White House. I think we saw every room, with the president sharing the history of every detail. I was really impressed.

After the tour, we gathered back in the Oval Office. One of the children asked, "Don't you have a dog?" The president smiled and said yes and abruptly called out, "Barney!" A wall opened and a big man in a bulging suit and sunglasses with ear buds ran in a few seconds later, holding a Scottish terrier in his arms. He deposited the hound on the rug and the kids were full of joy, chasing the little guy onto the South Lawn, Secret Service agent jogging in trail. They ran that dog as far as we could see, and then back again, disapproving Barney followed by kids followed by agents J and K. By this point, the dog was breathing so hard I thought it would die, and another agent appeared, scooped him up, and whisked poor Barney away through a different door in a different wall. Our visit was over shortly after that, and the entire group, regardless of political affiliation, was astounded by how much personal time and attention the president had spent with us. He genuinely cared about the space program and was personally engaged in trying to give whatever comfort he could to the families who were suffering so badly. It was a lesson in leadership for me.

Opportunities to visit the White House have continued after my time at NASA ended. I was invited to speak at a meeting of the National Space Council (NSC) in 2018, where they were discussing human exploration policy. Ironically, it was the same meeting where President Trump announced that there would be a Space Force. Even more ironically, before that meeting, as our group of attendees was walking through the security gate to enter the White House, a lobbyist friend of mine tapped me on the shoulder and whispered, "Terry, a bird just crapped on your head." Great, who actually has this happen to them? My buddy assured me that this was a sign of good luck. I was walking next to an Air Force three-star general, who was also a former F-16 pilot, and I grabbed him and said, "Sir, I need a wingman." We found a restroom, and he grabbed some wet paper towels and cleaned up my noggin.

I was immensely thankful and walked out into the waiting room where some of the cabinet and the chairman of the Joint Chiefs of Staff were mingling. While talking with the chairman, the most senior uniformed officer in the military, I called him "Mr. Secretary," because I thought he was Jim Mattis, the Secretary of Defense. When he corrected me and said, "I'm not the Secretary, I'm the CJCS," I about died from embarrassment. I told him it was the biggest faux pas of my career, but it was his fault because he didn't have a nametag on (generals don't have to wear nametags because everyone is supposed to know their name). He had a good laugh, and we went on to have a great conversation. What a beginning to my big day, when I was about to speak at the White House!

The council was discussing NASA's plans to create Gateway, a mini space station in orbit around the Moon. I had been vocally opposed to this plan because it would make getting to the Moon more expensive and time-consuming and was fundamentally a self-licking ice cream cone, conceived to provide a raison d'être for some very large NASA programs. Coincidentally, the congressmen whose districts stood to gain billions of dollars for these projects were its biggest proponents. I had even written several op-eds opposing it before the NSC meeting. Nonetheless, the White House invited me, knowing that I did not approve of NASA's plans. This really impressed me; in Washington you don't often see government organizations or politicians willing to have, much less encourage, public dissent, but there I was, telling the vice president and much of the cabinet that I thought this was a bad idea.

After the meeting, I must have gotten a "thank you!" from a hundred folks in all corners of the aerospace industry: NASA insiders, corporate executives, DoD officials. Everyone seemed to agree that Gateway was a bad and expensive and inefficient plan, but nobody would say that publicly because their business or contract depended on it. The Gateway plan has become so large that nearly every contractor at NASA has a stake in it. So onward we go, pursuing a plan that is a bad idea but nobody can stop, because congressional representatives and senators need to bring home the bacon and contractors need to keep their contracts. Nonetheless, it was a fun trip to the White House, bird poop and faux pas notwithstanding.

As astronauts, we hold a special position in the eyes of many, not only with respect to space policy, but throughout the nation and even the world. Many people hold astronauts in high regard, and in a universe (pun intended) that is usually full of spiteful, partisan politics, space exploration and those of us who have been fortunate to go there can be a breath of fresh air. Most of my colleagues dreaded their postflight Hill visits, but I really enjoyed them. I have gotten to know some extremely smart, experienced, and dedicated staffers working on various committees or at the White House, underpaid and overworked. But DC is also populated by some folks who aren't smart, experienced, or dedicated when it comes to space policy, and often they make capricious decisions based on bringing home that bacon. That's no way to run a space agency. It needs to be about the rocket science, not about the political science.

SPACE TOURISM

What You Need to Know Before Signing Up

The desire to travel into space, to visit the stars, probably predates *Homo sapiens*. I am sure that our Neanderthal ancestors looked to the heavens and wondered what it would be like to be a star voyager. An astronaut. In the near future, such travel is about to become much more accessible. Companies like Blue Origin and Virgin Galactic will be offering rocket flights into space, on suborbital missions that will give their passengers a few minutes of weightlessness and a view of the curved arc of our planet and her thin blue atmosphere. A company called World View will be offering balloon flights to the upper reaches of our atmosphere, where you will see the blackness of space and the curve of our planet. These experiences will cost hundreds of thousands of dollars initially, though hopefully they will eventually drop to the five-figure range. Which means they will be accessible to millions of people across the globe, unlike the few tourists who have purchased rides to the ISS for tens of millions of dollars. A golden age of space tourism is about to unfold.

Many of you (maybe most, by this point in the book?) would love to take the giant leap of flying into space. And I'm sure that some of you will, if only for a few minutes. So here is my advice for you.

First, take the meds. You will only be in space for a short time, so the odds of getting sick are low, but don't risk it. Take the medication that your flight surgeon recommends. This experience is so precious that you don't want it ruined by feeling like you're going to barf. I think that a very large percentage of average people without high-performance jet fighter experience would feel nauseous if exposed to weightlessness if only for a few minutes, so don't think twice. Better living through chemistry. That's my motto.

Second, don't focus on photography. You will only have a few minutes to experience the sensation of weightlessness and take in the view of our planet in a way that you never have before. Soak it in. Make it stick in your mind. Take brain pictures. Let the onboard cameras film the view; I am sure that these companies will have an array of high-def cameras rolling the entire time. Get a good picture of you with the Earth in the background, properly exposed and in focus. This is mandatory. Get a few of those. And then enjoy the few minutes of being in space without worrying about other photos. It will all end *much* quicker than you think. You will be able to watch the footage that the onboard cameras capture for the rest of your life, but make sure you sear those images into your brain. Your psyche. Your soul.

Think about floating before you go. When I was in Air Force pilot training, we used to chair-fly the next day's flight so that we would be ready for whatever our instructor would throw at us. We would sit in our office, a printout of the T-37 or T-38 cockpit in front of us, and think through every step and every procedure and every maneuver that we would fly, from engine start to acrobatics to engine shutdown. In the same way, chair-fly your spaceflight. Think of what takeoff will be like. Of what will happen during countdown. Of what it will be like to experience high g or zero g. Of what it will be like to float around. To exercise restraint and not flail your arms and legs around when it feels like you are falling. To do a few practice weightless maneuvers, gently touching the wall in front of you and feeling your body quickly move away. Of floating something to one of your fellow crewmates.

There are so many experiences that you will have in those precious few minutes—think about how you will experience them and do them all in your mind over and over before you fly, so that you can record them in your brain during those few precious minutes in space, capturing them to share with your friends and family. That is preparation that you will not regret.

Before launch, take time to get to know your fellow crewmates. Have a special dinner the night before launch. Make a mission patch that you all share and give away to friends and family. Fly a bag of those patches in space; they will be precious gifts to give away for years to come. Set aside a special time the night before launch to make toasts and celebrate the amazing experience that you will have the next day.

If the space launch companies offer different locations, consider the time of year and the view that you will have. Blue Origin and Virgin Galactic will initially launch from Texas and New Mexico, but if they eventually offer other locations around the world, think of the view you want to have. Do you want to see the desert southwest of the United States, which is spectacularly beautiful? Or maybe the deserts of the Middle East? If they ever offer a night aurora view, I would highly recommend it. They may eventually launch from Norway or Finland in December or January, and on a night when the sun is active, the view of the northern lights would be spectacular. I'm not sure what the future holds for these companies, but this is something to consider. No matter where you launch from, I can promise you that the experience of seeing your home planet below you, while you float in space, will be spectacular.

This chapter could not have been written just a few years ago, but trips like these will be a very real possibility in the very near future. I hope that many people are able to enjoy the sublime experience of spaceflight, if only for a few minutes. However, before getting overly excited, there is one thing that everyone needs to be aware of, no matter if they are launching in a rocket, spaceplane, or balloon. And that is—space is hard. And dangerous. These tourist flights will not be without risk, and if you decide to do it, there is less than a 100 percent probability that you will return. If you can live with that possibility, and more important, if your family will support you, then go for it.

Flying in space is an experience that very few humans have ever enjoyed. I promise you that it will be unforgettable. Start saving up now. And remember one thing: Take the meds.

ARE WE ALONE?
IS THERE A GOD?
AND OTHER MINUTIAE

My Take on Some Minor Questions

There are a few questions that everyone has about space travel, beyond the "How do you go to the bathroom in space?" and "What is it like?" They are some of the deepest questions that humans have ever pondered, questions that have confounded and informed and guided and challenged philosophers for centuries: Is there a God? Are there intelligent beings out there? I'm not sure why many believe that astronauts are particularly qualified to answer these questions; I for one do not have a PhD in religion or biochemistry, or anything for that matter. However, I did spend a fair amount of time off our planet and was maybe somehow closer to God and/or aliens, so that qualifies me to a small extent. I'll give it a shot.

First, the easy one. Is there a God? Maybe my insight will settle this matter once and for all and unite humans from all faiths around the globe! Or I can simply offer my own insights and perspective. I will address this from my own point of view, from my own experiences, and from what I've found to be true. I'll be as scientific as possible and not speak of religious views but simply share my observations about life and the universe.

One of the most important things I did in space was perform science investigations on my own body. Ultrasound scans of my brain, heart, and eyes. Laser and infrared measurements of my eyes. Constant measurement of my cardiopulmonary function, my weight, changes in height and dimension, the list goes on. I performed experiments on plants, worms, tissue samples, and rodents, learning more about biology and the human body than I ever imagined. And what I learned was unequivocal. Every single experiment I did pointed to a creator.

Let's take for example a wineglass. Imagine setting a pile of silicon on a rock in the mountains and waiting a billion years. In fact, let's have a billion piles of silicon, all sitting there for a billion years. Let there be wind, lightning, rain, storms, radiation, snow, and ice. Anything you can imagine happening in nature. At the end of those billion years, there would not be a wineglass. The random processes of nature could never create something as simple as a wineglass. Now imagine how much more infinitely complex an organism such as a single-cell amoeba is than a wineglass. It can reproduce, has DNA, has a variety of cell organs that convert solar energy into chemical energy, etc. Some of these organisms can even move and respond to their environment.

If something as simple as a glass wouldn't randomly create itself, how can a simple single-cell organism create itself? I was recently told that every living cell is comprised of between millions and trillions of molecules. As a scientifically minded individual, I just don't see any plausible scenario where lightning and wind and cosmic radiation would suddenly organize that many molecules from the primordial soup of water and carbon-based molecules and amino acids into a single simple cell. A living being, capable of reproducing with unimaginably complex DNA. Again, I'm no PhD or biochemist or evolutionary biologist. But common sense and general scientific knowledge tell me that couldn't happen without some help. Another thing I've noticed is that something as simple as cleaning my garage never happens on its own. Disorder in my daily life never turns to order without my effort. Things tend to degenerate to a more basic state, not a more organized one. It's the nature of how the universe works.

Because of these basic observations, I don't think that life would spontaneously happen without a creator. A wineglass wouldn't. A watch wouldn't. A clean garage wouldn't. And something as remarkable and complex as life certainly wouldn't. In my simple, fighter-pilot brain, the existence of life is de facto proof that there is a creator. *Res ipsa loquitur.* The thing speaks for itself. I'm not suggesting that there aren't natural laws that regulate life and evolution; of course there are. I'm simply saying that science leads me to conclude that someone had to set things in motion, to implement the laws of nature, and to create life. It doesn't make sense, from a scientific point of view, that life could happen without someone very smart behind it.

Beyond biology, consider the physical universe. From the ultrasmall to the ultrahuge, it is absolutely amazing. On the small end of the scale, physicists are constantly trying to find smaller and smaller subatomic particles. String theory is a popular idea in modern physics that describes the smallest of small things. This theory posits that there is a limit to how small matter can be, and it is 10^{-33} m. That's small. A 1 with 33 zeros to the left of it, *before* you get to the decimal point. Let's write that out: 0.000000000000000000000000000000001 meter. That's the smallest possible size particle, according to this well-accepted theory. Particles

Humans have been studying the natural world for millennia, and we have barely scratched the surface of ultimate scientific knowledge.

on that scale are made of strings, which when combined make larger particles, then even bigger ones, ultimately forming the protons and neutrons and electrons that we all studied in high school. Those make atoms, which make molecules, which make the things that physicists call matter. On larger scales, matter forms planets, then stars and solar systems, then star clusters, galaxies, galaxy clusters, and then, well, the universe. Which is 4.4×10^{26} m big. 4,400,000,000,000,000,000,000,000,000 meters. That's big.

So, the physical universe ranges from the nearly infinitely small (10^{-33} m) to the nearly infinitely large (4.4×10^{26} m). That's an astonishing range of sizes. And there are physical laws that determine precisely how everything works, on all scales. At the subatomic level, the weak and strong nuclear forces determine how atoms and molecules are held together. The electromagnetic force drives the ways stars are formed and burn their fuel over lifetimes that span billions of years. Magnetic fields shape solar systems and entire galaxies, funneling massive amounts of energy from the nuclear fusion of stars and the unimaginable violence and destruction from black holes. Earth's own magnetic field traps billions of tons of charged particles, violently ejected from our own sun, to form a "force field" around our planet, protecting life like you and me from deadly radiation, both from the sun as well as from ultrahigh energy particles from across the galaxy. It also funnels those charged particles down to our north and south magnetic poles, where electrons collide with the highest and most tenuous reaches of our atmosphere, creating an astonishingly beautiful and otherworldly river of green and red plasma flowing by the Earth's poles, the auroras—one of the most beautiful sights I have ever seen.

And then there is gravity, which holds it all together. Imagine that gravity was a little stronger. Not only would we humans be shorter, but our Earth would be zipping around the sun faster. Suns would burn their nuclear fuel at a faster rate, resulting in a vastly reduced life span, potentially precluding the possibility of life. Galaxies themselves might be smaller, rotating faster, producing more intense X-ray and gamma-ray bursts that would also make life impossible. All of this extra fuel-burning and orbital racing around would potentially drastically shorten the life of the universe—there might have been a Big Bang, then a brief period when stars and galaxies formed and burned brightly, followed by them quickly extinguishing and vanishing into eternal darkness, the universe populated by the ashes and cinders of dead, burned-out stars and flat-pancake galaxies. All because the gravitational constant was a little stronger than it is today.

It is fascinating to "what if" these questions. What if the weak nuclear force was a little weaker; would we even have molecules? What if the speed of light wasn't constant; would time travel be possible? What if the H_2O molecule did not expand in solid form, like nearly every other form of matter, and ice sank rather than floated? Life as we know it wouldn't be possible. What if the life span of stars, and the rate at which they burned their fuel, converting hydrogen into helium, carbon, iron, etc., was slightly different? What if the periodic table had only ten types of atoms on Earth, instead of more than 100? It would be a pretty boring and uninhabitable planet.

These "what if" questions are endlessly fascinating. You see, the universe is precisely balanced in ways that we haven't yet imagined. Humans have been studying the natural world for millennia, and we have barely scratched the surface of ultimate scientific knowledge. It seems that most of the universe is made of something called dark matter and dark energy, and we are just recently learning of its existence, much less what it is or what it is made of. In fact, one of the most important experiments on the ISS is called AMS-2, which is trying to help understand how much dark matter and dark energy is out there.

You see, folks, from my commonsense point of view, the physical as well as the biological worlds are so precisely tuned as to require a creator. In my humble, fighter-pilot opinion, there must be a *very* smart being out there who designed it all, who set it all in motion, with the precise laws of nature set in place to allow this amazingly beautiful and wondrous universe to exist and evolve as it does.

All of this begs the question: Are we alone? A question that has a few straightforward answers. It's also the question asked of astronauts more than any other, with the possible exception of "How do you go to the bathroom in space?"

My first answer is, "There are billions of planets out there, which would seem to imply that there is other life in the universe." NASA has recently launched several planet-hunting telescopes, including *Kepler* and *TESS*, which have found thousands of planets around other stars relatively near our own sun. When you extrapolate that number to the rest of our galaxy, you can safely assume that there are billions of planets in the Milky Way. Then extrapolate that to the billions of galaxies in the universe, and you can safely assume that there are a lot of planets out there. It would therefore seem to follow that if there is life on one planet, there must be life on many others.

However, as I mentioned earlier, life is so complicated that I don't think it would simply create itself. Which brings us full circle to the question of aliens. Although there are countless planets out there, and you would think there's life on some of them, I think that science requires that someone created it. Notice I'm not using the word "believe," I'm using the word "think." Let's keep belief out of this mind exercise.

At the end of the day, however, the debate over aliens is a moot point because of the sheer distance to those other planets. The closest star system,

Proxima Centauri, is roughly four light-years from Earth. It is also a star system with three suns orbiting one another, making life there all but impossible because of the continual erratic and extreme climate changes that any planet there would undergo, as portrayed in Cixin Liu's novel *The Three-Body Problem*. However, let's say for the sake of argument that there happened to be extraterrestrial life at this nearest star system. How long would it take us to get there and say hi?

The fastest velocity ever obtained by a human-launched satellite was by *Helios B* in 1976, one of a pair of NASA probes designed to study the sun. At its closest approach to the sun, it reached a speed of 240,000 km/hr. Assuming it could maintain that velocity all the way to Proxima Centauri—which it can't, because as it departed the solar system, the sun's gravity would slow it down by *a lot*—it would take 19,000 years to reach the nearest star. Let's take a more realistic speed—the velocity of the Voyager satellites, humanity's most distant probes to date, and their pedestrian 60,000 km/hr. The trip to Proxima Centauri would take 76,000 years. Ouch. We aren't going there anytime soon. Plus, those satellites are the size of a car. Now imagine the number of rockets and amount of rocket fuel that would be required to launch something the size of an aircraft carrier into space, and then accelerate it to those same speeds. That will never happen with current rocket technology. Even if it did, it would take as long as *Homo sapiens* have been on Earth to get to the nearest, uninhabitable solar system.

Let's say we develop an electric propulsion system that enables a vehicle to travel at a much faster speed; engines like this have been used for decades on a small scale. If we were able to build one on a much larger scale, such as the VASIMR engine developed by my good friend and fellow astronaut Dr. Franklin Chang-Díaz, a space vehicle might achieve a velocity of roughly 400,000 km/hr. It is still a 10,000-year trip, one way. That's 350 human generations. We know almost nothing about humans who were around 10,000 years ago.

There are other, more exotic forms of in-space propulsion: nuclear electric rockets, solar sails, even photon rockets that use light as their propellant. Let's be wildly optimistic and assume we can develop an exotic engine that could propel us at one-tenth the speed of light. This is *way* beyond anything that is remotely possible or feasible today. But let's assume that eventually happens. The nearest star is more than four light-years away. Our intrepid space voyagers would need time to accelerate to that incredible speed, spending half their trip accelerating and half decelerating. It would probably take at least 100 years. One way. Using technology that isn't even remotely possible on any serious rocket scientist's drawing board.

And these 100 years, 1,000 years, 100,000 years, whatever the duration, would be spent entirely in interstellar space. No resupply ships. No shielding from galactic cosmic radiation. Crop failure? You have the rest of your life to figure that one out. Meteor strike? You have the rest of your life to plug the hole with materials that you have with you, because Houston won't

> **We aren't even remotely close to being able to send a signal into space that would have a chance of being heard over the noise that is generated by our sun, and presumably the same is true of other intelligent beings.**

help. Communication with Earth? At first it will take minutes, then hours for a transmission. But a few hundred years into the trip, and a call home will take many years for someone on the other end to answer. Whatever problems you may have—technical, moral, political, spiritual—will have to be solved by the crew. Actually, it won't really be a crew. That group of humans who first set off for a new star system will be a group of pilgrims, in the truest sense of the word. Like Cortés when he arrived in Mexico, those pilgrims will essentially be burning the ships. When they leave the comforts of Earth, they will be leaving forever. By the time they get to their destination, their descendants, and hundreds of generations of descendants after them, will be gone. They will never return to the planet where their species was conceived. If that moment ever happens, it will be the most epic tale in human history.

How does this relate to aliens? First, let's assume we are traveling to a star system because we know that some intelligent beings exist there. We would presumably have heard an electromagnetic signal from them to be certain of their existence. Humans have been sending radio waves out into the universe for more than a century now, through radio and television signals. In order for an alien civilization to communicate with us, however, it would have to discern that signal above the tremendous noise that our sun generates. We aren't even remotely close to being able to send a signal into space that would have a chance of being heard over the noise that is generated by our sun, and presumably the same is true of other intelligent beings. They would have to make radio waves more powerful than their own sun in order for us to discern them. Besides all of these technical reasons, based on some of the news being generated in recent years by our world's leaders, maybe it's best that extraterrestrial intelligence not receive our broadcasts.

So, is there a God? The scientist in me tells me there has to be, because the universe in general and life in specific is just too complicated and incredible for there not to be. We aren't random accidents or afterthoughts. I came away from my spaceflights with one main thought: I don't have enough faith to be an atheist. Are there aliens? I tend to think there are, though I do believe they would have to be created. Does it even matter? Those other planets out there are so far away that I don't think we will ever be able to communicate with or visit our cosmic brethren out there. The good news is that the xenomorphs from the movie *Alien* won't be popping out of our collective chests anytime soon. The bad news is we won't be hanging out with Chewbacca either.

WHAT DOES IT ALL MEAN?

The Big Picture

Living in space is a complicated thing. It is at once profoundly sublime and utterly mundane. Being a fighter pilot has given me a very strong, matter-of-fact personality; I rarely get excited or emotional about anything, which is a great characteristic when the emergency alarm goes off, but it makes human relationships difficult. Our ability to compartmentalize emotions as pilots is legendary, but I will try to put thoughts and feelings to paper.

There are so many experiences that comprise a spaceflight, but I believe there is one profound one that supersedes all others: the realization that you are not on your home planet anymore. That you are here, in space, and the Earth is down there, over there, not where you are. Everyone who has ever lived and will probably ever live was born and died down there on that planet, over there, and you are not there. There was something profoundly, well, profound in that realization. Although I like to think that I'm still the same down-to-earth person that I was before my time in space, I know that it changed my worldview in profound ways. Now that my spaceflights are a few years in the rearview mirror, I think they changed more than my worldview. They changed my soul.

The first moment that really changed my perspective came about five minutes into my first launch. *Endeavour* performed a roll maneuver, giving me a view of the East Coast at night, and in one glance I could see the corner of America where I had grown up. I wasn't prepared for that moment and what I would see; I just suddenly had a view out of my cockpit of the land where I was raised, and it hit me. America isn't as big as I thought it was; I had just seen half its population in one glance out of the window. That snapshot was the beginning of a transformation in my way of thinking that would ultimately continue for seven months.

Minutes later, after *Endeavour*'s engines had shut down and we were floating peacefully over the North Atlantic, I had my first view of the planet in daylight. It was an intensely blue sunrise, something I would eventually experience more than 3,000 times, but this first one was special. It was a shade of blue that I had never seen before, another profound moment only minutes into my first spaceflight. I had spent my entire life preparing for that moment. I had seen all the IMAX films, read all the books, seen all the astronaut photography. I'd even spent the past decade working with fellow astronauts and hearing their tales of spaceflight. And yet, in this first glimpse of daylight, I saw something that I was entirely unprepared for: a new shade of blue—intense, vivid, burning, all-encompassing. It was unexpectedly clear that space was going to be full of surprises. And that was the understatement of the century. My seven months off our planet would slowly and profoundly challenge my worldview in ways I had not imagined.

My fifth day in space was a big one. We were docked to the ISS and had begun the weeklong process of installing the final two modules of the orbiting outpost, through a series of spacewalks, robotic transfers, and lots of internal maintenance work. As I was going to sleep that night, I closed my eyes and saw my first white flash from cosmic radiation. It was something I would eventually experience hundreds of times, but that first time was both impressive and sobering. Although those white flashes were pretty, they also meant that my body was being subjected to intense and dangerous radiation, a kind that doesn't exist on our planet. Which made me think—the universe is inhospitable and cold and dark and wholly incompatible with life, with the exception of our blue planet, as far as we know. I had a new sense of thankfulness and appreciation for our home, drifting through space like a giant spaceship carrying the entirety of our species on a timeless journey. We should take care of it. There is no plan B; there is only plan A. Though I had always thought flying in space would be about rockets and planets and spacewalks, in the end it was about our home. When asked, "What is your favorite planet?" I now respond, "Earth!"

A few hours before that white flash, I was watching the Mediterranean glide by silently below at night, a few thunderstorms flashing in the distance, a quarter moon hanging slightly above the horizon. There were beautiful city lights throughout my field of view, mostly yellow, but some blue and white. And it struck me: During the daytime, you cannot really tell that humans are down there; an alien flying by our planet in daylight might not even realize that we are here. But at night, it's a much different story because of those city lights. It would be fascinating to have a time-lapse movie of Earth at night over the past thousand years, or even ten thousand, to put into perspective how quickly humans have changed the planet. If you made that a billion-year time lapse, it would be dark for the entire time up until the last frame of the video, where you would see a quick flash of our city lights. Our presence today is visible globally. Well, mostly. In some places, you see lots of lights, and in others you see only darkness, even if there are lots of people living in that darkness. On that fifth night, I realized that I wasn't seeing population in those city lights, but rather wealth.

This was a profound realization, something I had never heard any fellow astronauts mention. But it was obvious once I first noticed. In some places like the East Coast of America, Europe, or East Asia, there were lights everywhere. Lots of people + lots of lights = lots of wealth. In some regions, like the Middle East, there were not very many people but lots of lights nonetheless, indicating plenty of wealth. However, there are a few places on Earth, most notably Africa, where there are almost no city lights even though there are a billion people down there.

It was striking to see how people live their day-to-day lives from outer space. Nowhere was this more apparent than the Korean peninsula. South Korea is a vibrant place at night, and Seoul is one of the brightest cities on the planet. The rest of the country is brightly lit, and the ocean surrounding South Korea is full of fishing-boat lights. Across the Yalu River, northern China is full of bright, yellow cities at night. In between there is a giant sea of blackness, with one small dot of white light—Pyongyang. I was struck how some people live in light (South Korea) and some live in darkness (North Korea). Yet another profound realization.

Another thing really stuck out when seeing those city lights at night. Flying over the Mediterranean, I could see the entirety of the Middle East in front of me, from the Nile River delta on the right to Cyprus and ancient Turkey on the left, stretching out to the Saudi Peninsula, Iraq, and even a faint glow from Tehran and dust from the Zagros mountains in distant Persia. Directly in front of me was a small cluster of lights. Tel Aviv was the brightest, but also Jerusalem, Aman, Beirut, and Haifa, all tightly grouped together.

> Maybe it was a case of being hopelessly optimistic, but it was pretty obvious from my vantage point in space that there was no reason for the conflicts we have—in the Middle East or anywhere else. We are all crewmembers on this spaceship, and we may as well get along and work together.

I thought, wow, Israel is such a small country; why all the fuss? I had been there on several occasions and found it to be an amazingly beautiful and vibrant country, and I absolutely loved the people and Shabbat dinners. I have also traveled to many Arab countries, enjoying their architecture, cuisine, culture, and hospitality. Some of the friendliest people and most interesting places on Earth were right there in front of me, looking through *Endeavour*'s pilot window, right next to each other. Israel and the Arab world, together, down there on Earth. Maybe it was a case of being hopelessly optimistic, but it was pretty obvious from my vantage point in space that there was no reason for the conflicts we have—in the Middle East or anywhere else. We are all crewmembers on this spaceship, and we may as well get along and work together.

Our STS-130 crew came back to Earth on the fifteenth day of our mission, landing at the Kennedy Space Center. As the pilot, I was required to power down a lot of the systems after touchdown, and I was the last person to get out of *Endeavour*. Our crew did a walk-around of the orbiter, did our medical tests, met with senior managers, reunited with our families, and finally made it back to crew quarters, an astronaut "hotel" that has been in use ever since the 1960s. And like any business trip, when I got back to my room I plopped down on my bed and turned on CNN. I don't remember exactly what the announcer was talking about, but it was blah, blah, blah, and whatever meaningless scandal there was back in 2010. I watched for about thirty seconds, grabbed the remote, and turned it off. I literally could not take it anymore—it was like fingernails scratching a blackboard. A few hours prior I had been in space, looking down on our beautiful planet, thinking grand thoughts about how to end poverty and war and what our ultimate destiny in the universe would be, and here I was back home, being bombarded with meaningless noise.

Our lives are so cluttered with noise; we need to figure out how to unplug and let ourselves breathe and think. Social media, cable news, constant entertainment, work emails and texts—it really is difficult to "be still" in the modern world. I firmly believe that this constant stimulation is unhealthy, intellectually as well as emotionally. On a recent trip to Vienna, Austria, I had the chance to give a talk in the same hall in which Beethoven played one of his masterpieces for the first time. It struck me to compare the type of life that those masters lived a few centuries ago to the lives that we live today. They had time to think. If they wanted to know something, they had to read a book or research it; if they wanted to compose music, they had to sit down and write it with pen and paper. Of course, today we have tools that are infinitely more efficient, and I'm not saying that is a bad thing, but I do believe there is something to be said for being alone and reading or thinking or writing. There is a level of creativity and intellect that is tapped by doing things in the old way that we lose when consumed by modern technology, and I believe there is a level of genius that has disappeared from human society because we don't invest the intellectual capital required to develop it in the way scholars and artists did in centuries past.

The more serious consequence of today's hurried pace is the emotional toll that our unceasing plugged-in lifestyle exacts. The next time you are at the airport, or on a train, or at a restaurant, look at the people around you, and they will all be heads down on their phone. Heck, the next time you're driving down the highway, look at the drivers around you. The Air Force called this CPA (continuous partial attention), an appropriate term. I struggle with putting the dang phone down and focusing on people around me, which can cause serious relationship problems. Another problem is constantly looking for affirmation from your online "friends," many of whom you may have never met. "How many likes did this get?" "What are people saying about my post?" I'm no therapist, but I'm quite sure that this isn't healthy for self-esteem or emotional well-being. To paraphrase the scene from *Skyfall*, when Moneypenny is shaving James Bond with a straight razor, "Sometimes the old ways are best." Putting down your phone, picking up pen and paper and writing a letter, reading a book, or getting outside and being still in nature all beat running from meeting to meeting while typing on your iPhone.

> **And then you look down and see this oasis of beauty, our planet and home in a sea of hostile inhabitability. There is a much bigger picture, and we should not be distracted by the silliness that often fills our lives.**

We could probably all use more of these things and less tech. I never expected to learn that lesson from space, but that moment in crew quarters after landing the shuttle brought it home loud and clear. When you see the universe and its vastness from the vantage point of space, it is indescribably beautiful. I could use a bunch of adjectives, but they don't do the view justice. The universe is immense, beautiful, vast, cold, barren, black, and in so many ways beyond human comprehension. And then you look down and see this oasis of beauty, our planet and home in a sea of hostile inhabitability. There is a much bigger picture, and we should not be distracted by the silliness that often fills our lives. There are real and important issues that matter, like poverty, the environment, safety and security, the economy, our families, etc. I have resolved to be a better steward of my time, not wasting it on trivial matters of little importance.

There is another change that I've noticed in my outlook after leaving Earth for seven months. I'm less of a black-and-white guy. When I was younger, it was easy to see things from a very simple point of view; some people were wrong and some were right, and everything was black and white. That is a convenient view of the world and one that is comforting. If you can say "those are bad guys" or "these are good guys," it simplifies your worldview and makes it unnecessary to think critically or challenge your own perspective.

But that's not how my brain works now. In reality, the world is a complicated place, and though, of course, there is evil, and there are also good and self-less people, most situations involve some degree of gray rather than a purely black-and-white analysis. The ability to empathize and see and understand others' points of view is crucial. Now I tend to try to understand why others think and act the way they do, and if there is something to be negotiated, I try to make it a win-win, allowing both sides to benefit from the deal at hand. I'm not sure why a trip to space would cause me to think this way, but it is a shift in my thought process. And I think it's a positive step.

This newfound global perspective has helped me realize what dangerous times we live in. A bold new era where nations around the Earth are turning toward autocratic strongmen as their leaders. Where people are fracturing and becoming afraid of others. Where there is near universal pushback on the ideals that made the late twentieth century so successful—liberal democracies and free-market economies. In my mind it feels like 1928, and I want to prevent 1940 from happening.

I have seen what a uniting force space exploration can be, because people from around the world love space. Projects like Apollo and the International Space Station have brought us together in cooperation unlike anything else in history. I believe that this unifying power is needed as we stand at a historic fork in the road. Do we choose the path that divides us? Or the path that unites us? Helping to steer us in the direction of unity is my new mission in life. All this while so many nations seem to be choosing the other path.

Seeing our planet floating through the universe has made it impossible for me to pick partisan or ideological sides in political arguments. There are much bigger issues for us to confront if we are to have a brighter future. I hope the next phase of my life, through speaking and writing and media, will impact hearts and minds around the globe. As we focus on building bridges, and not walls.

Wearing PPE before PPE was cool. It is required when going into the US Lab's observation window for photography—they didn't want us sneezing on the million-dollar glass.

ISOLATION

Better on Earth or in Space?

This was not a chapter I expected to write. When I put together a list of all the topics I wanted to discuss in this book, it included a mix of some expected, some fun, and some unexpected topics. Isolation, which of course is part of living on a space station for 200 days, was not one of them. There are elements of how I dealt with being isolated throughout the fifty-one chapters in this book. But until 2020, this subject was not a regular part of life on Earth.

And now it is a stark reality for most every person on Earth.

With that in mind, and with full recognition of the gravity of the pandemic reality facing us all, I will treat this topic in the same way I usually treat serious, life-threatening, "this-might-not-turn-out-good" situations: I'll use a bit of humor.

My life has been transformed in the same way that so many other lives have—I'm isolated and contained in my Houston home without normal human contact, unable to travel, much of my business completely shut down (except writing extra chapters for my book!), and running low on toilet paper. All this led me to make an interesting comparison: Which situation is better— being stranded in space or quarantined on Earth?

Let's start with space. On my second mission, we had been on the International Space Station for five months and were set to come home in a few weeks when the Russian Progress cargo ship had, to borrow a term from SpaceX, an "anomaly." In other words, a catastrophic event completely destroyed the vehicle. While the Russian engineers and scientists conducted a safety investigation to ensure that it was safe to launch the next Soyuz capsule, and with it our replacement crew, we were told our mission would be extended indefinitely. Which meant we were stranded in space, low on supplies.

Let's begin with the first concern: food. There is probably nothing more important to the morale of an expedition, a military unit, or a space mission than food. And the food in space wasn't bad. It tasted pretty good in general, and as I mentioned earlier in the book, we were mostly happy with it. There was decent variety, especially with some European and Russian food available. An odd quirk was the American beef, which, like any good wine, had a year stamped on each package. There's nothing like eating a vintage three-year-old sirloin (like sipping a fine wine).

There are some similarities between food on Earth and in space. Canned soup and frozen dinners are pretty similar to the space versions. Some of the off-the-shelf items we had in space were, well, off the shelf of a local grocery store. Cookies, candies, olives, tuna packets, beef jerky, and other things were just the same as what you get on Earth.

There were some profound differences though. First, beverages. There's no carbonation in space . No Diet Coke for me, no beer, nothing that would fizz, because in weightlessness that would make a big mess. There was no fresh bread in space either. I didn't think I would miss it because I don't normally eat a lot of bread, but the first thing I ate after landing was a fresh chicken sandwich at the airport kiosk. I'll never forget how good that bread tasted. There are very few fresh fruits and vegetables in space as well. There's no fresh anything in space, for that matter, which made a big hole in my diet for those 200 days.

What's more, there are no restaurants in space. Sometimes I just craved a burger or fried chicken. Or something from a fine French restaurant. Or barbecue. Or Mexican food like we have everywhere in Houston. It is good to eat out, at least occasionally. A lack of restaurants probably helped me lose weight, but there are real psychological benefits to eating out beyond simply the food, even if all you can do is carry-out. There is something beneficial about getting out of the house, going out for a treat, or enjoying time with friends and family.

Food: advantage Earth

Next, let's talk about work. In space, my work day was always varied. There was something different to do every day—experiments, maintaining equipment, preparing for spacewalks, exercising, media outreach, unpacking cargo ships, medical exams, etc. And though my work days varied, there were common themes. We woke up at roughly the same time every day, had a kick-off meeting to start the day around 0730 GMT, and then met again to end the day around 1900. There was always time blocked off for lunch and exercise, Sundays were usually free, and so on, which lent a familiar ebb and flow to our daily schedule that was comforting if demanding.

On Earth during isolation, it can be tough to establish this sort of predictable routine. First, you need to be disciplined enough to set up and follow a daily schedule. Just starting every day with a routine (alarm, shower, make your bed, breakfast, etc.) is a great thing to do, even though not all of us have that self-discipline, myself included. I'm lucky to have an office in my new house that is devoted to work, but for a few years after leaving NASA I had to use a couch as my office (which had its own advantages). There is a monotony to being isolated in the same house, day after day, under Earth's stay-at-home orders without a clear schedule or program to guide the day's events.

Work: advantage space

One of the recommendations I've given people suffering through this isolation is to try something creative. Start that novel you've been wanting to write. Take up drawing or painting. Have an acting competition with your friends over Skype. Or try my personal favorite: photography. We all have smartphones, so learn their obscure photography features like live mode, time-lapse, and slo-mo. Play with lighting and composition. Learn Photoshop and other software. There are lots of photography skills you can learn while you're stuck in your house or apartment.

There are also plenty of things to photograph *on* Earth, but nothing compares with photographing Earth *from* space. For seven months, whenever I looked out the window, I never saw the same thing twice. And every one of those views was overwhelmingly beautiful—deep colors in the oceans or deserts, dark jungles, blindingly white clouds, intricate nighttime city lights, otherworldly auroras, power beyond imagination in thunderstorms and hurricanes. It was so spectacular that our cameras, the best Hollywood-quality still and video cameras available, could not properly capture a sunrise from space, or the unique colors of the curvature of the Earth in bright daylight, or the wonder of the southern lights as our spaceship flew through their giant green flowing river of plasma reaching hundreds of miles above the planet. Like an upper-deck home run, this one is a no-doubter for me. But it doesn't have to be photography you pursue—it can be anything that unleashes your creativity and gets your mind off being confined at home.

Photography: advantage space

Going to the bathroom is another interesting, unexpected, and complicated comparison. Generally speaking, gravity is nice. It keeps everything moving in the right direction. Earth toilets are simple, their design and construction are pretty well perfected, and they usually work without a problem. (Unless you're in the embarrassing situation of being on a first date and you clog the toilet at your date's place, but that's rare and a whole other story.)

I did have to repair the space toilet on the ISS several times, and it was a complex operation, taking a lot of time and work. It was also fairly easy to have a minor disaster if you didn't perform your toilet business properly, and quite messy, as I explained in detail in chapter 24.

But this whole discussion comes down to two words: toilet paper. Despite three cargo ships blowing up in an eight-month period, the ISS never ran low on the stuff. Down here on Earth, there's a virus going around and everyone on the planet feels compelled to buy all the TP and paper towels in sight. When we have a vaccine and the pandemic finally subsides, I predict that nobody will buy toilet paper for six months.

Seriously, people, stop hoarding TP. Work is shut down, malls are closed, nobody's flying, but they're not closing down the toilet paper factory!

Bathroom: advantage space

Entertainment was another key component of my space mission. I watched more TV in space than I do down here on Earth. While working out, it was a real joy to disconnect from the daily grind and watch uplinked TV shows, movies, or sports events or listen to *Car Talk*. We wouldn't exactly have enough bandwidth to stream Netflix, Amazon Prime, Hulu, or Disney Plus, but the entertainment options were pretty good.

Now that I'm on Earth, especially during the virus mess, I actually have all of those streaming options available. I also don't have 2.5 hours of daily scheduled exercise that doubled as TV time (the gym was closed for physical distancing). But it is still nice to turn on a show or series at the end of the day.

Like TV and radio, I had music uplinked while in space. Every week my support team at NASA would send up a few MP3 files of recorded music from my Pandora stations. But down here on Earth, I can actually listen to live Pandora, Apple Music, Spotify, or the local radio. You name it, I can get it!

Beyond digital entertainment, there are also books. They are near and dear to me as an author and presumably to you since you are reading this book! We didn't really have much of a paper book library (there is no storage for such things on the ISS), and though I could have had e-books uplinked, I chose not to.

Entertainment: advantage Earth

One of the best parts of being in space was sleeping there. It was the soundest sleep I've ever had in my life and just so cool to close my eyes, snugly wrapped in a sleeping bag, and float. For the first few months in space, every time I closed my eyes I felt as though I was pitching forward and rolling left because of my inner ear's unique vestibular system. Falling asleep at night was no different, but after I got used to that pitching and rolling sensation it was somehow comforting knowing that I was in my "sleep zone," tumbling in darkness, floating, often listening to music or recorded sounds from Earth.

I love sleeping on Earth, too! I love my bed, comforter, sheets—just everything about it. But one of the experiments I did for my spaceflight proved scientifically what I knew intuitively. I wore a device to measure the quality of my sleep in my bed on Earth and compared it to my sleeping bag in space. The results were clear.

Sleep: advantage space

The question of necessary supplies isn't as cut-and-dried as you might think. It's true that we are running low on some things during the pandemic, and for certain countries like India the situation is dire and life-threatening. But for the most part, the supply chain for critical items in many places hasn't broken down. Yet.

Just before my Soyuz launch, we lost an Orbital Cygnus cargo vehicle when it exploded on liftoff—and with it half of my clothes, some food supplies, critical equipment and experiments, and Samantha Cristoforetti's spacewalk gear. So that was not great. Five months later, a few weeks before we were to return to Earth, we lost the Russian Progress cargo vehicle, so supplies started to get critical. And yet again, right after I left the ISS we lost a SpaceX Dragon, the third resupply vehicle in an eight-month period. The supply situation on the ISS was critical at that point.

From my perspective, neither the supply situation during my Expedition 42/43 mission nor the current pandemic supply situation is good. Hopefully, the pandemic we're experiencing will clear up soon and we will return to a relatively normal situation. But in the interim, have I mentioned this yet?

Stop hoarding toilet paper!!!

Supplies: tie

If I'm adding correctly, the score is Earth 2, space 4, and tie 1. Space wins! It's better to be holed up in orbit than it is down here on Earth in this moment.

In all seriousness, the coronavirus situation has no historical precedent. This is the first time that all of humanity has faced a common enemy. There have been pandemics in the past, but this is the first one in the connected age. It has come with unimaginable challenges but also an opportunity to unite as humans.

Dealing with the virus itself is the immediate priority. It is a deadly disease and we need to do everything we can to slow its spread until a cure and a vaccine can be developed. Beyond dealing with the actual medical disaster, there is widespread psychological stress that many of us have been living with in the coronavirus era.

If I were to pass on one piece of serious, sober advice from my experience of being unexpectedly stranded in space, it would be to focus on your attitude. You can make it through anything with a positive outlook and a will to survive. My attitude during that stressful time was simple: The situation would eventually end, we would make it back to Earth, and I would have the rest of my life down here. There was nothing I could do to change my circumstances, so why worry about it? Just make the most of the extra time I had in space, and then when I eventually got home and life got back to normal, I could move on.

That attitude really helped me and our whole crew make the most of an uncertain and stressful situation. Our plight wasn't nearly as bad as the pandemic situation, but it can serve as a metaphor that can hopefully provide some context.

So do your best to keep a good attitude. This isolation will end (and maybe it already has!), the economy will recover, and your life will return to a modicum of normalcy, hopefully sooner rather than later, but in any case it will eventually. And until then, don't hoard toilet paper!

ACKNOWLEDGMENTS

This work was quite literally a lifetime in the making, and it is an impossible task to acknowledge all who have contributed to it. There is no more appropriate place to begin than by going back to Oakland Mills High School and my English teachers, Ms. Herman and Ms. Mitchell. Let me say right up front—I am sorry! I was the worst English student since Ferris Bueller. I literally read only one book, *Death of a Salesman*, from cover to cover during my whole high-school career, and that one because it was only about fifty pages long. All the other books I "read" in four years of "honors" English were via Cliffs Notes. I sucked as an English student, and if there were such a thing, I would have been voted "Least Likely to Write a Book." But thank the Lord, I was lost and now I am found—*How to Astronaut* is my second book! After my first book, *View From Above*, I wasn't satisfied; it felt as though anyone could write *one* book. It wasn't until my second book that I finally felt like a legitimate author. When I ran my first marathon, it was OK and I was happy that I finished without having a heart attack. But after my second, I hadn't simply run a marathon, I had become a legitimate marathoner. That was the feeling I had after finishing *How to Astronaut*. I'm finally an author, and Ms. Herman and Ms. Mitchell would be proud and shocked.

Most of all, I want to thank my family, both immediate and extended. Writing takes a lot of time, and time spent writing is time away from them. Hopefully the royalties from this book will pay for my nursing home one day, and you won't have to worry about me. Just kidding. I love you, and you were a big part of this work.

|||| ACKNOWLEDGMENTS ||||

Next, friends and colleagues who are too numerous to list. You have contributed in so many ways, since my early days at the Air Force Academy all the way through my post-NASA career. At the end of the day people matter above all else, and what I learned from the remarkable people of the US Air Force and NASA during my career will never be matched by what I gave them in return. Of course, not everything is rosy and perfect in life, and I've learned a few things along the way about how *not* to treat people. To those who taught me sometimes-painful lessons, as we used to joke in the fighter squadron, you weren't entirely useless—you served as a bad example. Thank you for showing me the path in life to avoid—something often more valuable than seeing the path to take.

A few months before beginning this book, I met a great American and fellow author named Clint Emerson. He recommended a book by a guy named King. You may have heard of him. Stephen King. So I picked up the first Stephen King book of my life (I'm not kidding)—*On Writing: A Memoir of the Craft*, and it was exactly what I needed as a new author. In it he emphasized the importance of cutting word count by saying things as succinctly as possible. Of avoiding adverbs—not "I lazily walked to the lake while wistfully recounting a story with my friend who was vigorously complaining about the gently falling snow in the deathly still air," but rather "I walked to the lake while talking with my buddy who was whining about the cold." Mr. King also motivated me to pick up a copy of *The Elements of Style*, by William Strunk and E. B. White, the tried and true manual of grammar and the English language. I never imagined I'd read a book about how to write, and certainly not Strunk and White, but they were both indispensable on my journey to becoming a proper author. Thank you, Stephen and Clint and Strunk and White, I owe you.

I needed help from some of my NASA colleagues when dealing with my CRS in order to remember the many details I had forgotten. Rick Cole was one of my NASA flight surgeons during my Expedition 42/43 space mission, and Julia Wells was my Crew Medical Officer trainer, and they both helped my recollection for the space medicine chapters. Josh Matthew was my crew training officer for both my shuttle flight and my long-duration station flight, and he was always a great help for "Hey Josh, what does this acronym mean?" or "Do we still do that procedure?" Alex and Faruq, thank you for making those spacewalks possible and safe. Beth Turner was my family support person, and while others were trying to make life difficult, Beth made it tolerable. I am forever grateful to you for the support you gave me and my family.

To my STS-130 and Expedition 42/43 crewmates, you made my spaceflights safe, memorable, and experiences that I will always treasure as I attempt to share them with folks down here on Earth. All I can say is, more cowbell....

Of course, Don Pettit must be thanked, for both his friendship as well as his photography mentorship. In an office full of smart people, Don is one of the smartest. More importantly, he is a great dude. He's a legitimate genius, but also a very practical hands-on guy who can seemingly fix or invent anything. Being in the office with you was a highlight of my sixteen years at NASA, Don.

Jack Stuster is the world's leading expert on expeditionary behavior, and his two-decade-long experiment, called Journals, has become the definitive work chronicling the psychological status of every long-duration NASA astronaut since the beginning of the space station program. Jack understands the mental and human requirements of lengthy expeditions more than anyone else on Earth, and his expertise is the key to our future endeavors exploring the far reaches of the solar system. He also showed amazing insight when he noticed that I was the first ever ISS astronaut to have an improved mood during the third quarter of his mission. Normally, that is when crews get down, and they improve toward the end, but I was unusual in that I improved at the midpoint of my Expedition 42/43 mission.

The highlight of my time in space was making the IMAX film *A Beautiful Planet*, directed by the late Toni Myers. Working with her, her director of photography James Neihouse, and space consultant and fellow astronaut Marsha Ivins was one of the most special experiences of my career. I have since begun a new filmmaking career, and I learned from the best. Thank you, Toni; we miss you terribly. And I promise, if I ever see aliens, I won't not film them just because it's not on the shot list!

This book would never have happened were it not for my agent, Geoffrey Jennings, who believed in me and was willing to pitch me to my excellent publisher, Workman. Without great partners in your agent and publisher, an author is up the creek (or stuck in orbit without a rocket engine for the deorbit burn). I gave Geoffrey some not-so-great ideas for a book, and he politely told me, "These are great, I recommend you continue your day job as a speaker." When I brought him the idea for *How to Astronaut* he jumped to work, taking it immediately to our publisher.

My publisher, Workman, has been absolutely wonderful to work with every step of the way. Suzie Bolotin, the publisher, supported the idea for this book from the start and never gave us a chance to even talk to any other publishers—that meant the world to me. Workman's CEO, Dan Reynolds, signed off on this project without ever meeting me, when I had only one book to date under my belt. I needed two editors to encourage me and help turn my chicken scratch (in a figurative, Microsoft Word kind of way) into this beautiful work: Bruce Tracy and John Meils, as well as a spectacular copyeditor, Claudia Sorsby. Thanks as well to Kate Karol, Barbara Peragine, and Galen Smith. I'm just a fighter pilot, and I needed your help to bring *HTA* together, making sure the nouns and verbs and (rarely) adverbs would be fun for the reader. Moving on to the fun part of the book, thanks to Janet Vicario for an amazing book design, and to Vaughn Andrews for a book cover that will keep this work flying off shelves for many years to come! Finally, to Rebecca Carlisle and Diana Griffin, thank you for suffering my endless ideas for marketing and book-tour stops. Without your work none of us would have a job!

Lani, you're the best.

The people responsible for getting me in shape for my flights, and then even more importantly getting me readapted back to Earth afterward, are known as ASCRs (NASA acronym for strength and conditioning trainers). Mine were Bruce Nieschwitz for Expedition 43 and Corey Twine, Christi Baker Keeler, Mark Guilliams, and Jamie Chauvin for STS-130 and general training. After 200 days in space I had lost 0.0 percent of my bone density, so they must have been doing something right.

One more thank-you. For the person whom FS calls, in her Australian accent, "race car, race car, race car," thank you. I wouldn't be alive without you.

This writing thing is a blast. I have enjoyed it much more than I ever thought possible, and I hope I have many more books in me, fiction and nonfiction. I also hope that you, the reader, enjoyed this book, laughed, and occasionally said, "Wow! I never knew that."

Terry Virts
March 2020
Locked down in quarantine
Houston, Texas, USA
Earth

INDEX

PHOTO CREDITS